THE CHANGING FACE OF
AERIAL
WARFARE

ABOUT THE AUTHOR

Anthony Tucker-Jones spent nearly twenty years in the British Intelligence community before establishing himself as a defence writer and military historian. He has written extensively on aspects of Second World War warfare, including *The Desert Air War 1940–1943*, *The Eastern Front Air War 1941–1945* and *The Normandy Air War 1944*.

THE CHANGING FACE OF

AERIAL WARFARE

1940-PRESENT DAY

ANTHONY TUCKER-JONES

The
History
Press

Front cover illustration: A Royal Air Force Reaper RPAS (Remotely Piloted Air System) at Kandahar Airfield in Afghanistan. Sergeant Ross Tilly (RAF)/MOD. (OGL v1.0)

First published 2018 as *Spitfire to Reaper: The Changing Face of Aerial Warfare, 1940–Present Day*
This paperback edition first published 2023

The History Press
97 St George's Place, Cheltenham,
Gloucestershire, GL50 3QB
www.thehistorypress.co.uk

British Library Cataloguing in Publication Data.
A catalogue record for this book is available from the British Library.

ISBN 978 1 80399 383 6

Typesetting and origination by The History Press
Printed and bound in Great Britain by TJ Books Limited, Padstow, Cornwall.

MIX
Paper from
responsible sources
FSC
www.fsc.org   FSC® C013056

Trees for LYfe

CONTENTS

HERR HITLER'S MESSERSCHMITT

Leutnant Johann Böhm was flying his Messerschmitt Bf 109 over Dover with three others when they were pounced on by Supermarine Spitfires of No 74 Squadron. Their formation scattered in a desperate bid to escape. Böhm dived down toward the Elham valley with a Spitfire piloted by Sergeant Tony Mould in hot pursuit. No matter what he did, Böhm could not escape his pursuer nor their blazing guns. In his combat report Sergeant Mould recounted the dramatic encounter:

> He immediately dived to ground level and used evasive tactics by flying along the valleys behind Dover and Folkestone, which only allowed me to fire short deflection bursts at him. After two of these bursts smoke or vapour came from the radiator beneath his port wing and other bursts appeared to enter the fuselage. He eventually landed with his wheels up as I fired my last burst at him in a field near Elham. The pilot was

apparently uninjured and I circled round him till he was taken prisoner.

A dazed Böhm, who had crash-landed on Bladbean Hill, clambered from his stricken aircraft having received a nasty blow to his head. His damaged grey-green camouflaged Bf 109 was adorned with a shield decorated with a comic crying bird with an umbrella under its wing. Two of the propeller props had been bent back by the impact of the hard landing. Böhm looked at the trail of devastation left behind as he had ploughed through a flock of sheep. A farmer was later to complain he lost ten ewes. The date was 8 July 1940 and Böhm's Messerschmitt from Jagdgeschwader 51 had the dubious honour of being the very first German fighter shot down over Britain. It was first blood to the Spitfire.

Across the English Channel, JG 51 and JG 26 deployed in the Pas de Calais had been placed under First World War fighter ace 'Uncle' Theo Osterkamp, known as Kanalkampfführer or Channel Battle Leader. His job was to secure air superiority over the Straits of Dover and prevent British convoys using the Channel by attacking shipping with bombers. For the first time in early July 1940 German bombers had started flying inland on armed reconnaissance missions and by hiding in the cloud some had even reached the London area. The Battle of Britain had begun.

From the Luftwaffe's Eagle Day massed attack on Britain to NATO's assault on Gaddafi's Libya, aerial warfare has changed almost beyond recognition. The piston engine has been replaced by the jet and the pilot in some cases completely replaced by the microchip. Carpet bombing became global positioning system and laser pinpointed strikes using precision-guided munitions. Whereas a bomber's greatest enemies were once fighters and flak, these threats morphed into air-to-air and surface-to-air missiles delivered from beyond visual range.

Nonetheless, The Few of RAF Fighter Command continue to capture the popular imagination and admiration of successive generations. The idea of these gallant young fighter pilots rising up into the skies over southern England in 1940 to save the nation is compelling. The Spitfire and its cousin the Hawker Hurricane became national icons of this desperate struggle. The fact that Adolf Hitler never really intended to invade Britain is immaterial to their incredible courage. They showed that Britain would not be cowed and that the Luftwaffe could be defeated. American pilots did exactly the same at Midway when they proved the Imperial Japanese Navy was not invincible and decisively avenged Japan's surprise air attack on Pearl Harbor.

The development of the aircraft into a weapon of war occurred during the First World War with the emergence of the fighter and bomber derived from early reconnaissance biplanes. This gave rise to the very first fighter aces, who became national heroes. Men such as Theo Osterkamp, who claimed thirty-two kills. However, the conflict came to a close before the aircraft could be developed into a truly war-winning instrument. By mid-July 1940 Osterkamp had claimed six victories, making him one of the few pilots to achieve air-to-air kills in both world wars.

The nature of the air war changed during the Second World War as did the public's perception of what it could achieve. After the Battle of Britain the 'Bomber Barons' such as 'Bomber Harris' and General Spaatz became the focus with their campaign to bring Hitler's Germany to its knees by bombing his weapons factories into oblivion. This gave rise to the notion that bombers could win wars. Instead, Hitler's factories were eventually overrun. The cruel irony was that Germany would have run out of raw materials and manpower before the efforts of the bombers ever began to have a real effect on weapons production.

In the meantime, the crews of RAF Bomber Command in their Lancasters risked their lives night after night, while the crews of the United States Army Air Force in their Flying Fortresses did the same by day. Their combined sacrifice was enormous, the results questionable. However, they took the war to Hitler at a time when there was no Second Front and the Allies were still struggling in North Africa and the Mediterranean. Likewise, RAF Coastal Command took the fight to the enemy during the Battle of the Atlantic, when the U-boat menace sought to strangle Britain.

At sea during the Second World War a fleet's role was to protect its carriers so that their naval aircraft could attack enemy ships. The epitome of this was the Battle of Midway, fought by American and Japanese carriers in the Pacific. This single engagement was a turning point. However, by the late twentieth century a carrier's role was increasingly to conduct littoral warfare or coastal warfare, whereby a carrier's aircraft attacked land targets in the support of an army. The Korean, Vietnam, Falklands, Gulf and Balkan wars were prime examples of how naval air power had become subordinate to the ground war.

There is a general view that the air war in 1940–45 was a very crude affair and that it was not until the Korean and Vietnam conflicts that pilots became reliant on more advanced technology. Actually, the Battle of Britain started a brains arms race. There was a rapid evolution in air war science that significantly impacted on the course of the Second World War. Most notably, Britain had an early warning radar system that greatly assisted RAF Fighter Command to intercept Hitler's bombers.

In turn, the Germans used a fairly sophisticated beam system to guide their bombers to their targets. They also employed both short- and long-range radars to counter the Allied bomber offensive. The latter's bombers were fitted with airborne radars to warn of approaching enemy fighters and detect enemy towns. Either side's night fighters made use of airborne radar to track their foes in the darkness. For every scientific measure the boffins came up with there were countermeasures. Radars and navigation aids had to be jammed or even better deceived. The Germans also developed flying bombs and ballistic rockets.

The role of massive air forces remained all pervasive throughout the Cold War. Both sides were armed to the teeth, with their bomber fleets poised to strike at a moment's notice. Stanley Kubrick's movie *Dr Strangelove or: How I Learned to Stop Worrying and Love the Bomb* epitomised the paranoia of the Cold War and the threat posed by long-range bombers carrying nuclear payloads. The film culminates in a B-52 bomber pilot riding his nuclear bomb rodeo style to its target. Fortunately, this never happened for real. However, the B-52 was to inflict appalling death and destruction in Laos and Cambodia using conventional bombs in a bid to cut the Ho Chi Minh trail during the Vietnam War. Similarly, it was later used to pulverise targets in Afghanistan and Iraq. The use of the helicopter as a weapon of war also came of age during the Vietnam conflict.

While the fundamental experience of combat pilots may be very similar, technology has increasingly moved to distance them from their mission. No longer do pilots shoot at each other using cannons; since the 1960s it has been via missiles that use targeting radar. The last air combat where pilots visually engaged each other with cannons was in Korea. The real impetus to make aerial warfare clinical and stand-off was the Vietnam War, where Second World War-style heavy bombers jostled alongside precision-guided munitions to deliver their old-fashioned 'iron' free-fall bombs. Thanks to the constant glare of the media, public opinion finally made carpet bombing unacceptable.

By the twenty-first century some pilots were often operating from air-conditioned offices tens of thousands of miles away while their unmanned aerial vehicles or drones did the dirty work. The driving factor behind the rise of the deadly Reaper and Predator and the so-called 'Drone Wars' was the war on terror and the hunt for 9/11 mastermind Osama bin Laden. These were designed to deliver bunker-busting missiles into terrorists' lairs with pinpoint accuracy. They are now the future of aerial warfare.

The aim of the game has become to kill your enemy both in the air and on the ground with clinical precision and as little collateral damage as possible – that euphemistic phrase for civilian casualties, the spectre of which has haunted every air force commander since the days of the Second World War and the terrible firestorms of Dresden and Hamburg. Inevitably, though, civilians still get caught in the crossfire or are mistakenly targeted.

Ever since the Second World War, argument has raged over whether air power alone can win wars. Bending an enemy's will with the application of only air power is no easy feat. The likes of Operation Linebacker and attempts to strangle communist forces in South Vietnam failed miserably. Likewise, attempts to force North Vietnam to the negotiating table by bombing Hanoi were only partially successful and simply fuelled hatred for America. While no-fly zones may have influenced the Serbs during the Balkan wars, they did little to curb Saddam Hussein in Iraq when it came to crushing a widespread rebellion against him. Air strikes did, though, help facilitate regime change in both Iraq and Libya. Governments around the world continue to resort to 'bombs away' as a forceful policy option when all else fails.

I

EAGLE DAY

In the clear blue skies over southern England in the summer of 1940, Royal Air Force and Luftwaffe pilots were locked in a deadly struggle that was eventually to spill back across, Europe, North Africa and the Balkans. Radar-assisted RAF Fighter Command took six minutes for its planes to intercept the lumbering twin-engine German bombers and their far more dangerous fighter escorts. One can only imagine the sense of sickening apprehension that must have filled the young pilots as they lounged in their easy chairs, waiting for the inevitable alarm that would cause them to dash across to their planes. They would then leap into the cockpit and hurtle up into the air to stop Hitler's air force from smashing Fighter Command's airfields and Britain's vulnerable cities.

From 10 July to 8 August 1940 the Luftwaffe tried to force the RAF to fight by attacking exposed British convoys in the English Channel. Then from 8 to 18 August it bombed airfields, radar stations and shipping. The Heinkel

He 111 bomber was in the forefront of Operation Adlertag (Eagle Day) on 13 August 1940, which was the opening of Alderangriff (Eagle Attack) designed to destroy the RAF once and for all. The Heinkels of bomber group KG 26 were tasked to attack RAF Dishforth, while KG 27 struck Bristol, Birkenhead and Liverpool. KG 53 attacked RAF North Weald and KG 55 hit Feltham, Plymouth and RAF Middle Wallop. Despite the size of the raids, Adlertag failed to crush Fighter Command.

In the air battle that was to follow Britain had one key technological advantage: early warning radar, known as Radio Direction Finding or RDF for short. This was RAF Fighter Command's eye in the sky. Edward Fennessy, a radar expert at the Air Ministry, knew the network was at risk – 'General Martini, the Luftwaffe Chief Signals Officer, had by this time a pretty shrewd idea that we had an RDF system operational and he had to argue very forcibly with Göring to allow the Luftwaffe to attack the RDF stations.'

These attacks started on 12 August, however it proved difficult to damage the RDF sites. 'The lattice masts can't be seen from the air,' noted Ernest Clark, a wireless operator at one of the stations, 'and they were so designed that they could stand up on any two of their gimbals. And the blast used to go through them.' Only Hitler's Stuka dive bombers were able to do any real damage.

The death toll amongst pilots and aircrew during the Second World War was appalling. Survival during the 1939–45 air war often depended on lightning reflexes plus the mechanical reliability and robustness of the aircraft. It not only pitted man against man and man against machine, but also man against the elements, and once out of the aircraft the chances of survival were slim to say the least. On the ground or at sea, wounded soldiers and sailors at least had a fighting chance.

For the pilots and their aircrew fighting this deadly dual in the sky, the odds were simply stacked against them. Under these hugely dangerous conditions the basic instinct for survival takes over in crisis situations and enables an individual to escape death or face their fate. On 15 August 1940 Flying Officer Roland Beamont, with 87 Squadron, RAF, found himself in the thick of it:

> I fired at a Ju 87 [Stuka dive bomber] at point blank range, and I hit it. I don't know what happened to it. But I could see my tracers going into it. Then I came under attack from directly ahead and below. It turned out to be a Me 110 [twin engine fighter], doing a zoom climb straight up at me, firing as he came. He missed me. I rolled away from him straight behind another of his mates, a 110. I fired a long burst at him and his port engine stopped and started to stream smoke and fire, and pulled away from me.

For a fighter pilot, his battle for survival was not just a matter of skill operating his plane, but also intuition. His greatest fear was not so much being shot at or shot down, but rather failing to bale out in time, the failure of his parachute or, possibly worst of all, being burned alive trapped in the cockpit or whilst dangling from the parachute. Flying Officer Hugh 'Cocky' Dundas was over Folkestone at 7.15 p.m. on 22 August 1940 looking for an unidentified aircraft when his Spitfire span out of control. He recalled after being able to regain control desperately trying to open the hood:

> I stood up on the seat and pushed the top half of my body out of the cockpit. Pressed hard against the fuselage, half in, half out, I struggled in a nightmare of fear and confusion to drop clear, but could not do so. I managed to get back into the cockpit, aware that the ground was very close. A few seconds more, and we would be into it. Try again; try the other side. Up, over – and out. I slithered along the fuselage and felt myself falling free.
>
> Seconds after my parachute opened, I saw the Spitfire hit and explode in a field below.

Squadron Leader Tom Gleave, 253 Squadron, found himself in just such a terrible situation, which tragically seems to have been an all too common occurrence amongst fighter pilots. On 31 August 1940, he was returning to Kenley when his Hurricane was attacked from behind by a Jagdgeschwader Messerschmitt Bf 109 fighter. His enemy swooped up behind him, and after a burst of machine gun fire, the squadron leader's instrument panel was shot out.

Perhaps far worse, the reserve gasoline tank, located between the panel and the engine, was also hit. Horrifically, Gleave was sprayed with some of the aircraft's 28 gallons of aviation fuel and his stricken aircraft burst into flame. Pilots were warned that if their clothing was soaked in fuel they must switch off the engine and leave the throttle open, as this would help minimise the danger posed by exhaust sparks.

Miraculously, Gleave managed to wrench his canopy back and bale out. As he plummeted earthwards he realised his clothes were burning, so he did not pull his ripcord for fear of his parachute also catching alight. He fell several thousand feet before finally opening his chute. Unfortunately, he had broken his flying goggles the previous day and, as the flames licked up his body, his face was burned, sticking his eyelids together. This prevented him seeing an approaching Messerschmitt, although thankfully he could hear it being chased off.

By the time he had hit the ground Tom had been terribly burned, his face, eyelids, legs, hands and the underside of his right arm and elbow being badly affected. Despite his severe injuries, Gleave had survived and as he sat waiting for help, he concluded he would need the attention of a doctor after all!

Later he described baling out of his blazing Hurricane as 'like the centre of a blow lamp nozzle'. He was dangerously ill in Orpington Hospital, Kent, when his frantic wife Beryl arrived. She asked, with true English understatement, 'What on earth have you been doing to yourself, darling?' Tom replied, 'Had a bit of a row with a German.'

Only the previous day he had shot down four Bf 109s, though the RAF had lost a total of twenty-five fighters with ten pilots killed. Tom Gleave, despite his injuries, could count himself lucky, as 31 August was a particularly bad day for the RAF. Its losses were the heaviest to date with thirty-nine aircraft shot down and fourteen irreplaceable pilots killed.

That same day Messerschmitt 109s and 110s also pounced on Squadron Leader Peter Townsend of 85 Squadron. An Me 110 riddled his aircraft with bullets and gasoline gushed into his cockpit. His Hurricane, trailing smoke, dived and by good fortune did not catch fire. Deciding that discretion was the better part of valour, he baled out with a wounded left foot. At RAF Hornchurch, 54 Squadron was caught taking off as some sixty bombs fell along the dispersal pens to the gasoline dump. The Spitfires of Flight Lieutenant A.C. Deere, Sergeant Davis and Pilot Officer E.F. Edsal were all caught in the blast, but while the aircraft were destroyed, the pilots escaped with their lives.

Flight Lieutenant Deere recalled his unpleasant experience of being bombed while still on the ground:

> On 31st August, I was held up taking off by a new pilot who'd got himself in the take-off lane – didn't know where to go. He delayed me. By the time I'd got him sorted out and around, I was last off, and caught the bombs – and was blown sky high … But I got away with it – we all got away with it. I got pretty badly concussed – my Spitfire was blown up. I finished on the airfield in a heap.

For some that day it was sheer ingenuity that saved their lives in these extreme and confused combat conditions. Flying Officer Jimmie Coward was piloting one of the few cannon-armed Spitfires over Cambridgeshire when disaster struck. Coward was leading his flight in an attack on a group of Dornier 17 bombers when his guns jammed and his plane juddered as something struck it. He remembered feeling a dull pain, 'Like a kick on the shin in a rugby football scrum' and to his horror saw, severed from his left leg save for a few bloody ligaments, his bare foot lying on the cockpit floor. His flying boot had been blown clean off.

Coward succeeded in escaping his stricken aircraft, but with his foot spinning from the torn muscles the pain drove him crazy. Desperately he pulled his ripcord at 20,000ft and his chute jerked him up as it opened. His blood, though, was spurting from the tibial arteries. To make matters worse, the slipstream had sucked off his gloves, so his frozen hands could not shift the clamping parachute harness to reach the first aid kit in his breast pocket.

Most men would have passed out from shock. Coward, using the radiotelephone cable in his flying helmet, lifted his damaged leg and bound the thigh to stem the loss of precious blood. He drifted back over Duxford airfield (home to 19 Squadron) and was rushed to Cambridge hospital. Coward lost his leg below the knee, but not his life.

Fellow Spitfire pilot Desmond Sheen was almost not so lucky. He had a very close shave and it was only his presence of mind that saved his life over the aerial battlefield of southern England. Having fainted over his control column because of a leg wound, he came to and found his aircraft hurtling towards the ground at 500mph. At such a speed there was no way he could level out. In an instant he was sucked from the cockpit through the open hood and ended up straddling the fuselage. With his feet trapped by the top of the windscreen, he only just managed to leap free in time.

Flight Lieutenant Robert Stanford Tuck, of 257 Squadron, also had a similar close shave. His Spitfire's fuel tanks were ruptured, covering him in hot black oil. Luckily, he was not burned and parachuted down at Plovers, the estate of Lord Cornwallis, who had him escorted to the bathtub stating, 'Drop in for a bath any time, old boy.'

William 'Ace' Hodgson, a Hurricane pilot, risked life and limb by staying with his blazing plane. The reason for his selfless gallantry was that down below him were the Shell Oil Company tanks at Thames Haven on the Thames estuary. What would have happened if his Hurricane had crashed amongst them one can only guess. Hodgson switched off his engine to keep the flames in check and successfully made a dangerous wheels-up landing in Essex.

The constant nagging fear for RAF pilots during the Battle of Britain was undoubtedly fire. Whereas Luftwaffe pilots reached the English coast with their fuel tanks almost exhausted, the Hurricanes and Spitfires were often fully laden. This gave them a longer but potentially danger-ous combat period. Flying Officer Richard Hillary, 603 Squadron based at Hornchurch under Squadron Leader George Denholm, suffered a similar fate to that of Squadron Leader Gleave on 3 September 1940.

Having pursued a Messerschmitt Bf 109 fighter for too long, Hillary was attacked from behind. His Spitfire burst into flames and for several nightmare seconds he was trapped in the blazing cockpit. Fortunately his aircraft broke up, throwing him clear to open his chute. Advice to fighter pilots stated, 'If you are on fire DON'T open the hood until the last moment, as it will draw flames into the cockpit.' This, of course, was easier said than done when in a burning aircraft.

Sergeant D. Fopp, 17 Squadron, was also burned on 3 September. He attacked a large formation of Dornier 17 bombers from the unfavourable head-on position and found himself engaging their fighter escorts. Fopp suc-ceeded in scattering three attacking 109s despite having run out of ammunition, but one of them got below and behind him. A burst of German rounds punched through his plane, hit his radio and Fopp found himself sitting in a ball of orange fire. He quickly baled out and managed to extinguish his smouldering tunic and trousers. Many RAF pilots and Luftwaffe aircrew were not so fortunate.

Spitfire pilot Sergeant Jimmy Corbin, with 66 Squadron at RAF Kenley, made an important decision:

> By early September it had become clear the squadron was taking a hammering. Two days previously Peter King ... had successfully baled out of his Spitfire only to find that his parachute wouldn't open and he plummeted to his death. It was stories like these that led me to promise myself that I would never bale out even if it meant going down with the plane.

The tough Polish volunteers fighting with the RAF rapidly gained an aggressive reputation. Squadron Leader Zdislaw Krasnodębski of the Polish 303 Squadron (stationed at Northolt – the other Polish 302 Squadron was at Church Fenton) had a very remarkable escape. His squadron engaged a huge formation of German bombers and their escorts on 6 September 1940. With the sun in their eyes, the Poles recklessly attacked on the climb, never a very favourable position.

In the following aerial melee, Krasnodębski's Hurricane was hit. Shattered glass showered his instrument panel, face and hands. Furthermore, the petrol tank was holed and gasoline slopped into his cockpit with fire spreading quickly. Over Farnborough, Kent, Krasnodębski jumped out, but with 100 planes around him there was the extreme danger of being hit in the crossfire, so he fell 10,000ft before pulling his ripcord. This act saved his life. Krasnodębski's legs were smouldering and were beginning to glow. Had he pulled at 20,000ft the fire would have spread up his body to the chute lines. When he hit the earth the flames had only reached his knees, though it was to be a whole year before he flew again.

German fighter pilots endured exactly the same deadly experiences as their RAF counterparts. In one instance, fate conspired to add a tragic twist to the survival of Oberstleutnant Hassel von Wedel. At Hanns Farm, Bilsington, above the Romney Marshes, Alice Daw was getting her 4-year-old daughter Vera ready for a family outing. Her husband, William, was in the barn getting their car ready. Way above them were the rattling sounds of battle and the telltale vapour trails of accelerating aircraft.

Wedel, flying a Messerschmitt Bf 109, was hit at 6,000ft over Maidstone by a pursuing Hurricane. Struggling with his controls, Wedel felt his plane suddenly plunge from the sky at great speed and he crashed into the roof of Daw's barn. William Daw was knocked unconscious by the impact, whilst poor Alice, running from their cottage, had her skull fatally fractured by debris and little Vera was tragically killed outright. By some miracle, Wedel was hurled uninjured from his smashed aircraft.

The local fire brigade found him almost in tears in a pile of manure. All he could say over and over again was, 'I've killed a woman.' A kindly fireman went to the cottage to make Wedel a cup of tea and no one had the heart to tell him about the dead child.

2

LUFTWAFFE LOSSES

During the Battle of Britain what pilots needed from their fighters was agility, speed and armament that packed a punch. If the combination was right then it produced a deadly machine. This meant that both British and German aircraft were constantly fine-tuned and upgraded. By the start of the Battle of Britain the Bf 109E was armed with two 20mm cannon and two machine guns. The initial Hurricane which appeared in the late 1930s was equipped with eight machine guns, the subsequent model built in 1941 was up-gunned with two 20mm cannon and two machine guns. Likewise, the Spitfire had eight machine guns until they were supplemented by 20mm cannon. The German fighter had a fuel injection so did not lose power in a steep dive. In contrast, the engines on British fighters would cut out due to the carburettor being flooded with fuel, so had to be rapidly modified.

Such work did not often go smoothly. Fred Roberts was an armourer with the ground crew of 19 Squadron at Duxford, which was the first to receive the Spitfire in 1938. When the squadron gained some of the experimental Spitfire Mk IBs armed with 20mm Hispano cannon in late June 1940 no one knew how to maintain the guns and they kept jamming. Roberts recalled:

> We still had some eight-gun Spitfires on the Squadron, which was fortunate because the cannon stoppages seemed unsolvable. We took a lot of stick from the pilots over the stoppages. For a while, they wanted to blame the armourers for the trouble and then, when a full magazine of 20mm ammunition was expended, the pilots complained they only had six seconds of firing time against eighteen seconds with the old Browning guns.

On 31 August 1940, 19 Squadron lost three Spitfires as a result of cannon stoppages. Two of the pilots baled out but one was killed crash-landing his aircraft.

The Luftwaffe's Heinkel He 111 medium bomber became a very familiar sight to Londoners. This aircraft was the main type used in most of the raids against Britain in the summer of 1940. At the start of the Second World War the He 111 bore the brunt of the Luftwaffe's tactical bombing campaigns in Poland in 1939, Norway and Denmark in April 1940, France and the Low Countries in May 1940 and then against Britain in July–August 1940. Like the Dornier Do 17, the He 111 by 1940 was already facing obsolescence. It was too slow against modern fighters and slightly slower than the Do 17 and less manoeuvrable. However, it could carry twice the bomb load. As a result, a total of six German bomber groups (KG 1, 4, 26, 27, 53 and 55) equipped with the He 111 were involved in the battle.

On 22 February 1940 two fighters from RAF Drem, East Lothian, including a Spitfire armed with cannon, almost captured an He 111P intact. It was intercepted just before midday off St Abb's Head, damaged and the rear gunner wounded. The pilot managed to make a forced landing at Coldingham but after the crew got clear they set fire to it before they could be stopped. This act deprived the RAF of valuable intelligence.

Although the He 111 acted as a reliable workhorse for the Luftwaffe in the early Blitzkrieg campaigns, in the Battle of Britain it proved vulnerable to the RAF's agile fighters. The German bombers' radius of operation in daylight was limited by the distance their fighter escort could cover. Although an excellent fighter, the Bf 109 was designed for close support not long-range escort duties. Its pilots had just minutes in the combat zone before they were forced to fly back across the English Channel. The twin-engine Me 110, designed primarily as a long-range escort, was almost totally outclassed by the RAF's Hurricanes and Spitfires. Ironically, the Me 110 had to rely on the 109s or fly defensive circles when engaged, leaving the bombers to fend for themselves.

In an effort to stave off RAF Fighter Command, the He 111 and other German bombers flew in very tight formations to provide mutually covering fire with their machine guns. Luftwaffe pilot Ernest Wedding recalled, 'I flew my Heinkel 111 bomber in formation and I had to keep to my station. Even when British fighters started attacking me, I couldn't do any intricate manoeuvres within the formation or else I would crash into the other bombers ... A bomber pilot had to be as steady as a bus driver.'

During July 1940 the Luftwaffe lost thirty-two aircraft and three damaged. The following month eighty-nine were shot down and another fifteen damaged. Squadron Leader Sandy Johnstone, commanding 602 Squadron, noticed a change in German tactics on 31 August:

> Findlay led the squadron on a patrol over Biggin Hill and Gravesend this afternoon and tangled with a bunch of Ju 88s [bombers] and Me 109s all mixed together. This was unusual, for the escorts normally fly well above their charges. However it didn't stop the boys from claiming three 109s and a Ju 88, although Sergeant Elcome got shot up and had to make an emergency landing at Ford.

The Luftwaffe launched its first massed raid on 7 September 1940. Squadron Leader Johnstone was amongst those scrambled to intercept them:

> I nearly jumped out of my cockpit. Ahead and above, a veritable armada of German aircraft was heading for London, staffel [squadron] after staffel for as far as the eye could see, with an untold number of escorting fighters in attendance. I have never seen so many aircraft in the air all at one time. It was awe-inspiring ...
>
> They spotted us at once and, before we had time to turn and face them, a batch of 109s swooped down and made us scatter, whereupon the sky exploded into a seething cauldron of aeroplanes, swerving, dodging, diving in and out of vapour trails and the smoke of battle.

Pilot Officer Tom Neil, with 249 Squadron, had been on two fruitless patrols that day when on his third late in the afternoon he recalled:

> We were in the area of Maidstone and at 18,000 feet when we sighted the tell-tale ack-ack bursts and immediately after, an armada of Huns: a thin wedge of Heinkel IIIs, then Dornier 17s, all beneath a veritable cloud of fighters, 109s and 110s.
>
> The cry went up, 'Tallyho!' After which we slanted towards them purposefully. Twelve of us against 100, at least. Dear God! Where did we start?

Mary Smith, daughter of the village postmaster, in Elham, Kent, recorded succinctly in her diary on 7 September, 'Terrible. Attack on Hawkinge in the morning. Masses of raiders over 5–6pm. Terrible night attack on London.' That day Flight Officer Crelin 'Bogle' Bodie, 66 Squadron, after tangling with Bf 109s, decided not to abandon his stricken aircraft:

> I heard the bullets strike the side of the kite, but when I opened the throttle to try and escape the machine didn't respond. Then suddenly she seized up altogether. The propeller stopped dead with me 15,000ft in the air. I knew then I had no choice but to try and land the plane quickly.

He made a successful belly-up landing in a farmer's field.

The Telegraph's special correspondent, Harry Flower, was at Lympne in the Folkestone area on 11 September and watched as the RAF pursued the retreating Luftwaffe after another raid on London:

> A dozen Heinkels and Dorniers were heading back seawards in half a dozen different directions. On the tail of each was a Hurricane or Spitfire. And then they began to fall. Three Dorniers with tell-tale wisps of smoke showing from their engines as they came lower dived for the sea in the hope of reaching the opposite coast ... A great Heinkel passed over my head flying low and in obvious distress following a rattle of machine gun fire up in the sun. A Hurricane slipped over some tree-tops on a hill, poured a short burst into the Heinkel, which caused the bomber to lurch wildly and then 'hedge hop' across field after field trying to find a safe landing.

Piloted by Heinz Friedrich, the bomber crash-landed on Romney Marsh. Four crew, including Heinz, climbed out carrying a fifth after setting their aircraft alight. Harry Flower and his photographer got an exclusive, including a photograph of a Spitfire circling the burning bomber.

By the end of the year Flight Lieutenant Peter Brothers, 257 Squadron, was feeling relieved. 'The sporadic raids went on into December, when they reverted to 109s carrying bombs. ... The winter was coming and clearly there was not going to be an invasion. The Battle of Britain had been won.'

Heinkel losses during the Battle of Britain forced the Luftwaffe to have a serious rethink and the bomber was switched to night operations and a variety of specialised support roles. Along with the Ju 52 transport aircraft, the He 111 also found itself bearing the burden of resupply operations on the Eastern Front. Most notably they were used to throw a lifeline to the German army trapped at Stalingrad between November 1942 and February 1943. Almost 200 were lost attempting to ferry ammunition and supplies into the German pocket. By the end of the war the He 111 had been relegated almost solely to a transport role, its Blitzkrieg glory days long past.

The Spitfire, with its superior speed, climb rate, operational ceiling and range, inevitably stole the limelight. It could get to 10,000ft much quicker than the Hurricane. Due to the obscured visibility with the Hurricane's panelled canopy some pilots flew with it open. Neither fighter had any heating. Pilots felt the Spitfire was elegant while the Hurricane was more rugged. Certainly the latter was easier to maintain and the Spitfire was not without its faults. Taking off in the Spitfire required flying left-handed while the undercarriage was retracted, and the long nose made landing tricky.

Flight mechanic Joe Roddis, with 234 Squadron, said, 'Everyone thought the Spitfire was the most marvellous thing on wings but without the Hurricane, we'd have been in real trouble. There were twice as many Hurricanes in the Battle of Britain as there were Spitfires.' Pilot Officer Thomas Neil, 249 Squadron, noted with good humour, 'Every second German pilot who was shot down by a Hurricane will say that he was shot down by a Spitfire. That was the folklore that went on.'

In reality they both played an important part by complementing each other's different capabilities. Flying Officer Jeffrey Quill, 65 Squadron, reasoned:

I took the view myself that it took both planes to win the Battle of Britain. Neither would have succeed on its own because the Hurricanes required the Spitfire squadrons to attack the Messerschmitt 109s while the Hurricanes concentrated on the bombers. ... You sometimes hear people saying, 'The Spitfire won the Battle of Britain.' Well, that's absolute rubbish. The Spitfire and the Hurricane won the Battle of Britain.

On rare occasions not only did the crew survive being shot down, but the abandoned aircraft also made it to the ground intact. Just two German bombers brought down over England made safe landings after the crew had baled out. One such landing occurred during Hitler's 'Little Blitz' on the night of 23/24 February 1944, involving a Dornier 217M heavy night bomber (successor to the Dornier 17). Operation Steinbock (Ibex), or the 'Little Blitz', commenced on 21 January 1944 and was a futile Nazi retaliatory night air offensive against London.

Dornier 217M U5+DK was piloted by Oberfeld Wedel (Senior Warrant Officer) Herman Stemann, navigator, commandant and bomb aimer Unteroffizer (sergeant) Walter Rosendahl, radio operator and dorsal gun turret operator Unteroffizer Hans Behrens and under gunner Unteroffizer Richard Schwarzmuller. They had taken off from Melun airfield south of Paris with some 270 planes involved in the raid (the official British history states 161). Their attack was to bomb the large arc of the Thames with targets ranging from Croydon to the City of London.

By 1944 London was probably the best-defended city in the world and the Luftwaffe had its work cut out even more than in 1940. The bombers stood little chance of survival if intercepted by the RAF's Mosquito night fighter, with its fast-speed, target-hunting radar and deadly four 20mm cannon that could tear a bomber to shreds.

Rosendahl's bomber had developed a defect in its starboard engine on take-off. This resulted in him flying at some 15,000ft, 3,000ft below the prescribed altitude. As U5+DK approached London it came under anti-aircraft fire, so flew off course and dropped to 9,000ft. The plane was then hit and the illumination for the pilot's instruments went out.

Fearing they would not make it, the automatic pilot was engaged and the crew jumped out, parachuting safely in Wembley, Middlesex. The aircraft flew on, gradually losing height. Surprisingly the anti-aircraft batteries did little harm to the plane or its 2 tons of bombs. In the meantime, the raid did some damage, but many of the bombs proved to be duds (probably due to the slave labour employed in the Nazi munitions factories).

Fifty miles north, the air raid warning sounded over Cambridge. Mrs Jane Riglesford of 302 Milton Road hurried to her air raid shelter. When she re-emerged, to her absolute amazement she was confronted by U5+DK, which had made a perfect wheels-up landing in the allotments at the end of her garden. The baffled Home Guard could find no trace of the crew, whilst the bomber was found to contain 600 gallons of fuel, enough to have got it home, as well as three incendiary containers, one with 590 2¼lb bombs and two others with 140 incendiaries each. It was a miracle that the plane landed and an even greater one that it had not exploded on impact.

For Hitler this massed attack was not a success. The British claimed of 106 enemy bombers detected, only ninety crossed the English Channel, of which just fifteen managed to bomb Greater London. Of the 114 tons of bombs actually dropped, only 49 tons fell on Greater London. The Luftwaffe, operating under severe difficulties and shortages, launched only six more manned raids of any significance on London, the final one being on 18 April 1944.

This last major raid on the British capital spelt disaster for Heinkel He 177 6N+AK and its crew. On the 18th some twenty bombers of the crack KG 100 Group attacked Tower Bridge. The heavy He 177s had superior defensive armament in comparison to the British heavy bombers, with nearly 1 ton of plate armour protecting the crew, but they needed it. The bombers were to make a feint towards the Midlands first in an effort to confuse the RAF.

The crew of 6N+AK Werknr (Work No.) 2377 were Pilot Feldwebel (Sergeant) Heinz Reis, observer Feldwebel Winand Hock, radio operator Unteroffizer (Corporal) Johann Wehr, flight mechanic Unteroffizer Georg Speyerer, rear gunner Obergefreiter (Leading Aircraftsman) Werner Heidorn and rear dorsal turret Obergefreiter Friz Kopf. The bomber was carrying twelve SC250 bombs each weighing 550lb.

It was when 6N+AK made its turn over Newmarket that it was attacked by a dreaded Mosquito night fighter. On the Mosquito's second pass, Reis had his controls shot away. As the big bomber began to drop, he gave the order to abandon the aircraft. Hock was first to leave, then Wehr. Reis became trapped by one of his flying boots in the exit hatch as he tried to escape, but fortunately the huge plane in its death fall tilted to starboard and Reis pulled his foot from the boot before parachuting to safety. Tragically, Kopf's parachute shredded and he plummeted to his death.

Speyerer's body was found in the wreckage at Butler's farm, 1½ miles north-east of Saffron Walden in Essex. The unarmed bombs came down a mile or two to the north.

The German Blitz was never on the scale or intensity of Bomber Command and the USAAF's offensive against Germany's cities. Also it was not until 1944 with Operation Pointblank and the change in target priorities that Bomber Command actually began to significantly harm the German war effort. The air offensive has both its advocates and its detractors, but whatever the morality of the German and Allied air campaigns, the fact remains that the losses on both sides were appalling.

For every airman who survived the air war, thousands did not. The Battle of Britain, 10 July to 31 October 1940, was but a foretaste of things to come. In it RAF Fighter Command lost 503 pilots and 915 fighters. The Luftwaffe's losses were far greater; their bomber and fighter crews suffered 3,087 killed and lost 1,733 aircraft. In 1939–44 the Luftwaffe suffered a staggering total loss of almost 100,000 men (which included 60,102 killed, 21,301 wounded, 9,521 training fatalities and 5,993 training injured).

German losses in aircraft were colossal; in 1939–42 alone they lost well over 5,240 planes just in the West. By the end of the Normandy campaign in 1944 they had lost another 3,500 aircraft. During early 1945 the Luftwaffe had to endure even more futile bloodletting, defending Nazi Germany's dwindling airspace. It is interesting to note that Britain, America and Germany's overall air losses were broadly comparable.

3

STRIKING PEARL HARBOR

They came in steady waves, 181 Japanese fighters, dive bombers and torpedo planes roaring across Oahu's blue–green hills of Kahuku Point, Hawaii. It was 0745 on Sunday, 7 December 1941 and they were headed for the US fleet. At anchor inside Pearl Harbor were ninety-six warships of the US Pacific Fleet under Admiral Husband E. Kimmel. The Japanese attacked in four groups toward the harbour and the air bases of Ewa, Hickman Field and Kaneohe.

The Japanese had intended their declaration of war to arrive just before the strike on the American fleet – instead it arrived thirty-five minutes after the start of the attack. There was a short lull in the battle around 0830, when the first wave of attackers departed, which gave the Americans time to improve their defences. The result was that the second wave of 170 Japanese aircraft suffered more than double the number of planes shot down in the initial onslaught.

The Japanese were in the forefront of developing aircraft carrier warfare in the vast Pacific Ocean. Like the Americans, they believed that carriers were best used as strike platforms that could destroy enemy carrier forces as well as battle fleets. In the winter of 1941 Japan planned to obliterate the US Pacific Fleet's carriers: the USS *Lexington* and USS *Saratoga*, which had both been commissioned in 1927, and the newer addition, USS *Enterprise*, which had joined the fleet in 1938.

To achieve this under the code name Operation 'Z', six Japanese carriers were directed to attack Pearl Harbor. The strike force selected for the mission was Vice Admiral Chuichi Nagumo's 1st Air Fleet, consisting of the carriers *Akagi, Kaga, Hiryu, Soryu, Shokaku* and *Zuikaku*, the light cruiser *Abukuma* and nine destroyers supported by the battleships *Hiei* and *Kirishima* and the heavy cruisers *Tone* and *Chikuma*.

The carrier concept was not new. Just before the First World War both British and American pilots had flown aircraft off slightly modified warships. However, getting the plane back on the vessel when under way was almost impossible. Sea planes proved a much easier option as they could be retrieved from the water using a crane. Although the driving factor was over-the-horizon reconnaissance, before the First World War weapons trials were conducted with ship-launched aircraft. Seaplane carriers, though, proved too slow and it was not until July 1918 that the modified battle cruiser HMS *Furious* conducted the world's very first carrier strike against land targets at Tondern.

The first true carrier, HMS *Argus*, was commissioned in September 1918, but this was built from an unfinished liner.

It was the Japanese who produced the first purpose-built carrier in 1922 with the *Hosho*. This ship proved to be too small and was followed by the much larger *Akagi* and *Kaga*. America's first carrier, USS *Langley*, appeared the following year but like the *Argus* was a conversion using a collier. During the 1930s development of the carrier ensured they became large and powerful platforms for the projection of naval air power.

The massive raid on Pearl Harbor heralded a much wider war. Within hours of the attack, Japanese task forces were also en route for the American islands of Guam and Wake. Although Guam fell quickly, Wake proved a much tougher nut to crack. The Japanese simultaneously invaded Hong Kong, the Philippines and Malaya.

The US Pacific Fleet had been in the Hawaiian Islands since May 1940 with the aim of acting as a deterrent to Japanese aggression toward British and Dutch colonial interests in Southeast Asia. In some US naval circles it was thought to be the height of folly having the fleet so far forward, especially as the support facilities in the Hawaiian Islands were clearly inferior to those on the American West Coast.

Since June 1940 Hawaii had experienced three major security alerts as well as numerous air raid and anti-submarine drills. Yet, despite Japanese–American relations spiralling downwards, no preparation had been made for the possible outbreak of war. Two reports in March and August 1941 had specifically identified the threat of a surprise carrier air attack. Instead the US armed forces preferred to remain in a state of denial. General George C. Marshall, the US Army Chief of Staff, described Hawaii as 'the strongest fortress in the world'. Besides, intelligence indicated the Japanese were mustering their strength to attack British and Dutch possessions in Southeast Asia and the Philippines. A simultaneous attack on American interests in the Pacific was dismissed as highly unlikely.

On paper, Pearl Harbor's air defences were considerable; the US Navy and Marine Corps had 250 aircraft stationed on Oahu Island, while the Army had another 231 aircraft. Anti-aircraft guns numbered just over 200 (consisting of 82 x 3in AA guns, 20 x 37mm AA guns and 109 x .50 calibre machine guns); the target number was supposed to be double this with some 420 guns (98, 120 and 302 respectively). In August, a radar warning system had been established but it functioned on a part-time basis – this situation was aggravated by a turf war between the Signals Corps and the Army Air Corps. Comically, the radar only operated from 0400 to 0700.

Some of Hawaii's aircraft were more than capable of taking on the 360 Japanese aircraft about to be thrown at the island. However, squabbles over training versus patrolling, poor intelligence and commanders unwilling to disperse their aircraft for fear of saboteurs meant the aircraft were not in the air at the crucial moment.

In total, Hawaii was supposed to have 148 pursuit planes and by the end of March 1941 these included fifty Curtiss P-36 Hawk and fifty P-40 Warhawk fighters (both aircraft were combat-proven in the service of other countries). The F4F Wildcat, while it could not match the manoeuvrability of the Japanese 'Zero', had heavy firepower and could withstand the rigours of aerial combat. In contrast, Japanese aircraft were designed light in order to give them range and were not armoured; as a result they could not take too much punishment.

Further to the west, the Americans had been working on their defences at Wake Island, which was an important stop-off en route to the Philippines and a place from which to keep an eye on the Japanese-held Marshall Islands; Peale, which hosted a seaplane base; and Wilkes Island, which was to be a submarine base but work had yet to start.

Wake Island's defences were under Navy Commander Winfield Scott Cunningham. Resources for the physical protection of the island were woefully inadequate. These consisted of the First Marine Defence Battalion numbering some 450 men under Major P.S. Devereux. This unit was so under strength that only half of its twelve anti-aircraft guns could be manned at any one time. To compound matters, there were no fire control or early warning radars.

In the air over Wake the situation was little better. Major Paul A. Putnam's marine fighter squadron VFM-211 provided limited cover. The squadron was equipped with twelve F4F-3 Wildcats but, having only just received them, it was not familiar with this aircraft type. Nor were the protective bunkers for the Wildcats complete when the Japanese attacked.

Japan's 1st Air Fleet assembled in Tankan Bay in the Kurile Islands on 22 November 1941 and began to sail four days later. The Japanese Imperial Navy took great pains to hide its tracks and avoided the sea lanes frequented by merchant shipping. They also avoided passing near any American naval air installations from which reconnaissance aircraft operated. Although heavy seas and foul weather greeted the Japanese task force, it arrived off the coast of Oahu on the evening of 6 December 1941 without being detected. It then reached its attack positions some 275 miles north of Pearl Harbor at 0600 on the 7th. An hour later the first wave of planes were in flight to their targets.

In 1941 the air groups embarked on Japan's six big fleet carriers were made up of a mixture of fighters, dive bombers and level bomber/torpedo planes. For example, *Akagi*, flagship of the fleet carriers, embarked eighteen Mitsubishi A6M 'Zero' fighters, eighteen Aichi D3A 'Val' dive bombers and twenty-seven Nakajima B5N2 'Kate' bomber/torpedo planes. The three smaller carriers only generally had a fighter and bomber/torpedo aircraft combination.

The Nakajima B5N2 that entered service in 1939 was one of the principal aircraft involved in the Pearl Harbor attack; 144 of these torpedo bombers took part in the air strikes. The following year they delivered fatal blows to the US aircraft carriers *Hornet*, *Lexington* and *Yorktown*. Similarly, 126 Aichi D3A dive bombers were involved in the attack on Pearl Harbor.

The aircraft from the *Hiryu*, *Kaga*, *Soryu*, *Shokaku* and *Zuikaku* participated in the raid, with the *Akagi* acting as flagship. However, the inexperience of the aircrews of *Shokaku* and *Zuikaku* meant they were only given a supporting role with the bombing of the airfields on Oahu. Both these carriers were brand new and had only entered service with the fleet in that summer. On 7 December the radar station at Opana, near Kahuku Point, was still on at 0702 and picked up the Japanese planes 137 miles to the north. Unfortunately, the Army Aircraft Warning Service Information Center at Oahu dismissed them as a flight of B-17s due to arrive from the mainland.

On that fateful Sunday all the fleet's battleships, except the *Colorado* that was in dry dock on the West Coast, were moored in pairs at Pearl Harbor. Japanese pilots flying over them were amazed. Commander Fuchida Mitsuo recalled, 'I have never seen ships, even in the deepest peace, anchored at a distance less than 500 to 1,000 yards from each other ... this picture down there was hard to comprehend.' At that moment the flight commander radioed Admiral Nagumo, 'Tora, Tora, Tora!' This repeated code word, meaning 'tiger', stood for 'We have succeeded in surprise attack.' Two minutes later the torpedo bombers were amongst Battleship Row to the east of Ford Island. Japanese bombs penetrated the battleship *Arizona*'s forecastle and detonated the forward magazine. More than 80 per cent of her 1,500 man crew were killed or drowned. Three torpedoes hit the *Oklahoma* and she rolled over with 400 of her crew.

To make matters worse, lined up wingtip to wingtip to protect them against sabotage were nearly 400 US Army, Navy and Marine Corps aircraft. The planes at the main Army Air Corps airfields, Bellows, Wheeler and Hickman, were swiftly reduced to scrap and the hangar facilities damaged. At Ewa, the Marine Corps air station suffered the same fate. Japanese dive bombers destroyed most of the three squadrons of Catalina flying boats at the Navy seaplane base at Kaneohe Bay as well.

The second wave hit the *Pennsylvania*, which was in dry dock, causing minor damage. However, the two destroyers in the same dock, *Cassin* and *Downes*, were almost completely destroyed. The destroyer *Shaw*, in a nearby floating dock, had her bow blown off.

By an extraordinary stroke of good luck, the carriers *Enterprise* and *Lexington* with their escort cruisers and destroyers were delivering aircraft to Wake and Midway Islands. The *Lexington*'s sister ship *Saratoga* was at San Diego on the US West Coast undergoing a short refit. *Lexington* was subsequently lost at the Battle of the Coral Sea in 1942, but the *Enterprise* and *Saratoga* survived the war.

Of the eight American battleships caught by the Japanese, half consisting of the *Arizona, California, Oklahoma* and *West Virginia* were sunk, and the *Maryland, Nevada, Pennsylvania* and *Tennessee* were severely damaged. Also sunk were three destroyers and four smaller vessels, while three light cruisers and a seaplane tender were badly damaged.

The Americans also lost 188 aircraft with a further 128 damaged. Casualties amounted to 3,435 killed or wounded. The Japanese only lost twenty-nine planes with seventy damaged. Five P-36 Mohawks were able to take off during the attack and are credited with shooting down two Japanese Mitsubishi A6M2 'Zeros' for the loss of one P-36; these were among the first American aerial victories of the Second World War. At Ewa Marine Air Station and Kaneohe Naval Air Station not a single plane was capable of flight. Eventually, a few US Army and Navy fighters took off from Hickman field to look for the Japanese task force, but it had long gone.

Radioman 1st Class Raymond M. Tufteland, serving on the USS *Chicago*, sailed back into Pearl Harbor on 12 December:

Our force entered Pearl to witness a ghastly sight of sunken ships – oil covered water – wreckage and ruin. We first passed the Nevada which had been beached to prevent sinking. Next one was California – badly damaged and on the bottom. The hull of the Oklahoma then came in sight after having capsized. The Tennessee and

West Virginia behind her were both damaged. However the team got underway and left the West Virginia still on bottom. The Arizona was completely blown up and a twisted mass of iron.

Bodies were still being taken from ships and out of the water a week after the attack. It was a sight none of us like to remember but must be avenged!

The daring Japanese attack was a remarkable success, putting out of action most of the US Pacific Fleet and giving the Japanese a free hand in the Southwest Pacific. Crucially, though, they had missed the American carriers, the attack's prime targets; they also missed the oil tanks and other installations that would have greatly hampered the American recovery. Perhaps foolishly, Nagumo refused his air commanders' pleas for a third strike to eliminate the unprotected oil tank farms and maintenance facilities.

Pearl Harbor marked the dawn of full-blown carrier warfare. With America's battleships out of the way for at least six months, the US Pacific Fleet could only take the offensive with carrier protection. This led to the concept of the fast carrier task force with the carriers' dive bombers and torpedo bombers acting as substitutes for the battle-ships' 16in guns. Indeed, the following year the *Enterprise* and *Yorktown* got revenge when they sank the *Akagi*, *Kaga* and *Hiryu* at Midway.

4

MIDWAY TO VICTORY

The Battle of Midway was one of the greatest and most decisive naval engagements of the sprawling Pacific War. Thanks to superior intelligence and improved tactics, the Americans thwarted a Japanese fleet five times the size of their own forces. What followed was a deadly game of 'cat and mouse' involving carrier-based dive bombers that culminated in a significant American victory. In the space of just five minutes, Admiral Nagumo's flagship, the *Akagi*, as well as two other Japanese carriers, the *Kaga* and *Soryu*, were hit by aircraft from the American carriers *Yorktown* and *Enterprise*. It was a triumph for American naval air power.

Following the success of their attack on the American naval base at Pearl Harbor, the Japanese had started 1942 with very high hopes. Admiral Nagumo, with five large carriers, four fast battleships, three cruisers and nine destroyers fresh from the conquest of the East Indies, sailed from the Celebes Islands near Borneo through the

Straits of Malacca and into the Indian Ocean. At roughly the same time, Vice Admiral Ozawa, with another force consisting of a small aircraft carrier, six cruisers and eight destroyers, sailed from Malaya to attack shipping in the Bay of Bengal.

For a week Nagumo's task force roamed the Indian Ocean, ending the Royal Navy's 150-year dominance of the region. Nagumo's carrier planes bombed British bases at Colombo and Trincomalee in Ceylon (modern-day Sri Lanka), while Ozawa sank almost 100,000 tonnes of shipping in the Bay of Bengal in just five days. For Britain it was a humiliation; Admiral Sir James Somerville's fleet, with just two modern carriers holding fewer than 100 aircraft, avoided the Japanese. The heavy cruisers *Dorsetshire* and *Cornwall* plus the small fleet carrier *Hermes* were sunk. The rest of the British Eastern Fleet had little choice but to withdraw to East Africa.

While these operations were under way, the Japanese mulled over their strategic options. The Combined Fleet command proposed attacking Midway Island 1,100 miles west of Hawaii to draw the American fleet to them. Commander Miyo Tatsukichi, the air officer of the Operations Section in Naval General Staff, pointed out that the proposed New Guinea–New Caledonia–Fiji operations were in the enemy's backyard. He argued that at Midway the Americans would have the support of land-based planes as well as long-range bombers from Hawaii. In contrast, the Japanese would be far from the support of land-based aircraft, at Midway there would be no element of surprise and, even if they did capture the island, holding on to it would be difficult. Tatsukichi reasoned New Caledonia was a far better objective.

Then, on 18 April 1942, all argument between the Combined Fleet and Naval General Staff stopped when US Army Air Force B-25 bombers launched from the USS *Hornet* attacked Tokyo. All opposition to the Midway operation ceased. On 7 May the Japanese and American fleets converged in the Coral Sea and fought their first pitched battle. This was the very first time in history that a naval engagement was decided by aircraft and not by gun barrages. It was also the very first time that the main units of the opposing fleets never saw each other.

F4F Wildcat fighters, SBD-5 Dauntless dive bombers and TBD Devastator torpedo bombers from the carrier *Yorktown* played a major part in the battle, sinking the light carrier *Shoho* in an attack lasting just ten minutes. The Japanese had lost their first aircraft carrier. Next day, on 8 May, *Yorktown*'s dive bombers damaged the carriers *Shokaku* and *Zuikaku*. In response, Val dive bombers and Kate torpedo bombers attacked the *Yorktown*. A bomb scored a direct hit on the flight deck and pierced three decks before detonating. Numerous fires broke out and damage control parties fought to control them. Although damaged, *Yorktown* was able to get back to Pearl Harbor for repairs.

Dive bombers from the American carrier *Lexington* also attacked the *Zuikaku* and *Shokaku*, but only the latter was hit by a single bomb. In the meantime, the Japanese launched a counter-strike and managed to hit the *Lexington* with two torpedoes and two bombs. Although the resulting fires were brought under control, the vessel suffered a series of explosions as a result of leaking aviation fuel vapour and had to be scuttled. Quite remarkably, the *Yorktown* was back at sea after just four days. The repair teams had worked around the clock and the carrier was ready just in time for the battle of Midway.

Although both fleets did not come in sight of each other, they both lost a carrier each and had one damage. The battle was regarded as a tactical victory for the Japanese. The only compensation that the Americans could draw from the bloodletting was that they had stopped Japan's effort to cut the supply route from Hawaii and Australia by gaining a foothold in New Guinea. It also proved that the Japanese were not invincible.

While the Battle of the Coral Sea involved relatively modest naval forces, the impending confrontation at Midway was on an epic scale involving hundreds of ships and even more aircraft. Even as the fighting in the Coral Sea was still raging the Japanese were readying their strike on Midway. The Midway task force gathered off Japan on 5 May 1942 ready for a two-pronged attack.

The main part of the Japanese attacking force, consisting of almost 200 ships including eight carriers, eleven battle-ships and nearly 600 carrier-based planes, steamed toward Midway. A second smaller diversionary force consisting of two carriers and four warships headed for the Aleutians in the hope of luring the American fleet northwards and out of the way.

US Admiral Chester Nimitz, thanks to the cracking of the Japanese codes, knew of Japan's plans before the task force even reached Midway. Ignoring the activities of the Japanese Northern Force heading for the western end of the Aleutian Islands and the Second Carrier Striking Group making for the middle of the chain, the American fleet sailed on Midway.

Tackling the Japanese First Carrier Striking Force was a daunting task for Nimitz. He had no battleships and just three carriers, but he did know his opponents' every move. When the Japanese reached Midway on 4 June 1942 the American fleet was lying in wait just north of the island and out of range of Japanese carrier-based reconnaissance aircraft. Japanese aircraft flying bombing missions over Midway did spot some American ships but assumed the force was limited to several cruisers and destroyers. No precautions were taken to protect the Japanese carriers, nor were the raids on Midway restricted or stopped.

While almost all the Japanese carrier-based aircraft were over Midway, the first wave of American aircraft struck the Japanese fleet. Fortunately for the Japanese, their anti-aircraft gunners blasting away shot down thirty-five of the initial forty-one American planes sent to bomb them. However, the Japanese continued to ignore the warning signs; they could not conceive that there might be another carrier in the area, so no effort was made to recall their pilots or regroup.

The second wave of American aircraft arrived just minutes later and caught the Japanese napping. On 4 June at 1022 dive bombers from the *Enterprise* attacked *Akagi*. A 1,000lb bomb smashed through the deck into the hangar, igniting torpedo warheads and fractured fuel lines. A second 500lb bomb exploded amongst the aircraft on the flight deck. Within thirty minutes the blaze was completely out of control.

At 1026 seventeen Douglas SBD Dauntless dive bombers from the *Yorktown* attacked the *Soryu*, scoring three hits down the centreline of the flight deck. The first 1,000lb bomb exploded in the top hangar and took out the forward lift. The second bomb caught the strike aircraft lined up on the flight deck. The third penetrated down to the lower hangar and blew up between the centre and aft lifts.

The *Kaga* was caught by four bombs dropped by Dauntless dive bombers launched from the *Enterprise*. Five more bombs were near misses. The fully armed and fuelled aircraft on the deck exploded and the vessel's fuel lines were ruptured. After half an hour the crew abandoned ship, but the carrier burned for another nine hours until she blew up.

Nagumo transferred to the light cruiser *Nagara* and *Akagi* was eventually scuttled. The combination of bombed-up aircraft and severed fuel lines turned the *Soryu* into an inferno and she was abandoned after twenty minutes of desperately fighting the fire. The stricken carrier remained afloat for eight hours until her exploding magazines took her to the bottom of the sea.

The carrier *Hiryu* launched eighteen Kate torpedo bombers and nine Zeros in the dawn attack on Midway. The vessel then missed the fate of the other three carriers. At around midday her aircraft helped disable the *Yorktown* with three direct hits. At 1445 *Hiryu* followed up this attack and a torpedo delivered a mortal blow to the *Yorktown*. While this second wave of attacks was being conducted, the American carrier launched ten Dauntless dive bombers to find and punish the *Hiryu*.

Two aircraft from the *Enterprise* located the *Hiryu* and by 1600 twenty-four dive bombers including ten from the *Yorktown* were in the air. An hour later they found *Hiryu* about to launch a third strike against the *Yorktown* with nine aircraft. Four bombs hit the centreline of the carrier's flight deck. American B-17 bombers from Midway and Hawaii also strafed the *Hiryu* with machine gun fire. Although the forward part of the ship caught fire, she withdrew west. Soon, though, the fire spread uncontrollably and Japanese destroyers picked up the survivors. Although the vessel was scuttled, she refused to sink and remained afloat until 0900 on 5 June.

Yorktown was hit by three bombs but managed to continue operating her aircraft until hit with two torpedoes. On 6 June a Japanese submarine put two more torpedoes in her and the vessel capsized and sank. Further losses were inflicted on the Japanese fleet. Cruiser Division 7 had been sent to shell Midway ready for a Japanese landing. On the way the heavy cruiser *Mikuma* and the cruiser *Mogami* collided after manoeuvring to elude an American submarine. The *Mikuma* was lightly damaged but the *Mogami*'s bow was caved in. The pair were then attacked by eight B-17 bombers flying from Midway but their bombs missed. They were followed by a dozen dive bombers from Midway but they failed to achieve any direct hits. The following day dive bombers, this time from the *Enterprise* and *Hornet*, found and sank *Mikuma* and damaged the cruiser *Mogami* and two destroyers.

The Japanese defeat at Midway during what was essentially a purely carrier battle marked a major turning point in the Pacific War. The defeat gutted Japan's carrier force and the country could simply not match America's huge carrier-building programme. Only three more fleet carriers were ever completed by Japan. It was also the first significant defeat inflicted by the Allies on the Japanese.

From Midway onwards, the Japanese adopted a defensive posture in the Pacific, permitting the American and

Allied forces to seize the offensive. The Japanese Navy would suffer successive defeats in the Philippine Sea and Leyte Gulf that witnessed the loss of yet more aircraft carriers. In June 1944 the last of their naval aircrews were slaughtered in the 'Great Marianas Turkey Shoot', and by the time of Leyte they had hardly any pilots remaining.

In a desperate bid to slow the steady advance of US air power across the Pacific, the Japanese resorted to suicide dive bombers from late 1944. These were fighters and dive bombers packed with explosives, though purpose-built rocket-powered planes were also produced. In response to these attacks, Allied carrier planes began to patrol much further afield from their task forces. Anti-aircraft defences were also greatly increased.

The peak of these Kamikaze attacks came during the bitter Battle of Okinawa in April–June 1945, when they tried to destroy the American invasion fleet. Amongst the casualties was the carrier *Bunker Hill*, which was hit by two Kamikazes and suffered severe damage. However, no carriers were sunk or even put out of action. During the three-month campaign the Japanese lost 7,800 aircraft (more than 1,900 of which were suicide planes), while American carriers lost 539 planes.

For sailors being attacked by suicide bombers it was a truly terrifying ordeal. All they could do was hope their fighter cover and intense flak would protect them. Advance warning did little to help, according to Quartermaster Fred Poppe:

> Kamikazes just poured at us, again and again. It scared the shit out of us … We'd get warnings about half an hour before they appeared, and the waiting was scary

too, the knowing what was coming when those pilots' one wish in the world was to kill you.

Many ships were damaged or sunk but at great cost to the Japanese. The planes were allegedly piloted by volunteers but many were inexperienced young men who were cajoled into killing themselves in the defence of Japan. Most were shot from the sky before they ever reached their targets. Relative to the size of the US Navy, they did not inflict that much damage and ultimately did little to deter the Americans. Such tactics were repeated on 11 September 2001 in New York with terrible results.

5

SODOM & GOMORRAH

During the Second World War much of Germany's war industries had to function under constant air attack by the Allies' strategic bomber fleets, which saw the RAF striking by night and the USAAF by day. Until the German surrender they concentrated on the destruction of Germany's industrial cities, individual factories and vital lines of communication. However, the Allies' war against Adolf Hitler's armaments factories up until early 1944 has been described as haphazard and uncoordinated. It was not until they concentrated on Hitler's synthetic oil plants that his factories and armed forces began to grind to a halt, with fatal results.

The British Target Committee had first called for intelligence on the German ball bearing industry, which was vital for the production of not only aircraft but also tanks, in the autumn of 1941. The following year, British intelligence assessed that a Swedish company,

Svenska Kugellager Fabrik through its German subsidiary Vereinigte Kugellager Fabrik, with two main factories in Schweinfurt, provided the bulk of Germany's needs. Helpfully, the latter's principal rival, Kugelfischer, was also located at Schweinfurt. British intelligence also identified two other major plants at Steyr in Austria and at Carstadt-Stuttgart, Berlin. Along with thirty-five smaller plants, they supplied 75 per cent of German industrial needs; the rest came from Sweden, France and Italy.

The vital production of Hitler's weapons was placed in the very capable hands of Albert Speer, an architect by profession, who at the age of 36 was appointed Minister of Armament and War Production in February 1942. Speer proved to be exactly the man Hitler needed. Within a year, the Reich's supply of steel, coal, oil and other raw materials reached its highest levels ever. To Speer's credit, the production of weapons and munitions likewise reached remarkable levels. By making use of existing civilian industrial capacity, he was able to double monthly output during 1943.

The Allies were not ignorant of the important role being played by armament factories in the Nazi-occupied territories. Initially Britain attempted to disrupt France's motor transport production for the Wehrmacht, in particular the Renault factory at Billancourt that was producing 14,000 trucks a year. The RAF raided the site on 3/4 March 1942 with 235 bombers. They were still perfecting their techniques and used target illumination flares, followed by incendiary and high-explosive bombs. In total, 470 tons of bombs were dropped on Billancourt and the raid appeared a resounding success.

Air Marshal Sir Arthur Harris, Commander-in-Chief Bomber Command, who had selected Billancourt as his very first target, stated:

> It is somewhat ironic that the first completely successful operation carried out after I took command should have been not only a diversion from the main offensive against Germany but also a precision attack on a key factory. This was the attack on the Renault works, near Paris, one of the French factories most actively engaged in producing war equipment for the enemy. This was a very short range target and it was almost undefended, which meant that we could attack it in clear weather and brilliant moonlight and that aircraft could come down very low to identify the factory ... The Renault factory was high in the list of collaborating French factories which had been given to me and this extremely destructive raid not only deprived the enemy of a considerable quantity of equipment, but was also of some value in discouraging the production of war material for the enemy elsewhere in France.

Harris painted a rather overly optimistic picture of the attack, especially as all but twelve aircraft claimed to have bombed the target. Although a sizeable tonnage of bombs was dropped and the initial assessment was that their concentration around the aiming point was exceptional, the final judgement was a little different.

Subsequent bomb damage assessment conducted by the RAF indicated that not only had French civilian casualties been high, but also the plant suffered a loss of less than two months' production capacity. Harris attempted to 'discourage' other French factories with a series of bomber raids. The Gnome et Rhone works at Gennevilliers, and the Ford factory at Poissy, as well as the Dutch Philips works at Eindhoven, were all bombed that year with limited results.

Italian dictator Benito Mussolini's armament industries felt the wrath of the RAF's bombers in the autumn of 1942. To support the efforts to drive the Axis forces out of North Africa, Bomber Command pounded the industrial cities of northern Italy. Genoa suffered six nights of attacks, Turin endured seven nights and Milan was subjected to one daylight attack. The latter showed just how weak Italy's air defences were. A shaken Mussolini declared there would be a nightly evacuation of the industrial cities of the north. The RAF had tried to attack Milan as early as June 1940 from Marseilles, but despite Italian troops attacking French soil, France had almost given up the fight. When the bombers tried to take off, French trucks were deliberately driven onto the runway to obstruct them and the attack had to be aborted.

The USAAF also had a go at shutting down the Billancourt factory on 4 April 1943, when eighty-five bombers dropped 251 tons of bombs. The leading formation of eighteen aircraft placed 81 tons of explosives square on the target. Unfortunately, the follow-on force was not so accurate and again civilian casualties were incurred. Although the Germans failed to intercept the bombers en route to target, their fighters harassed them all the way back to Rouen. When the RAF Spitfire escort appeared, they were also attacked and lost five aircraft for the loss of no Luftwaffe fighters.

The attack seems to have been a success. Afterwards, when the French Resistance smuggled photographs of the

Renault works back to Britain, they showed the Americans had inflicted severe damage on the factory. Billancourt was clearly of considerable value to the Germans, who, using slave labour, took nine months to completely repair the damage. German efficiency was such that they even managed to increase production from 1,000 vehicles a month to 1,500. The USAAF returned in June to attack the Triqueville airfield and aircraft factory at Villacoublay, also in the Paris area. Again the Luftwaffe vigorously defended these targets and the Americans lost five aircraft.

The seizure of Czechoslovakia's armament factories was to prove a major long-term blessing for the Nazi war effort. They represented a 1,200-mile round trip for the RAF and USAAF. Such great distances meant that the bombers were exposed to German fighters for much longer. In the RAF raid on Pilsen on 17 April 1943 more than 28 per cent of the attacking force was put out of action.

Air Marshal Harris noted in his memoirs:

In April and May 1943, my Command made two attempts to destroy the Skoda Armament Works at Pilsen, which had become of much greater importance to the enemy now that Krupps at Essen had been so heavily damaged. Unfortunately both attacks were unsuccessful; the bombing concentrations were remarkably good for so distant a target, but the main force had aimed with great accuracy at markers placed a mile or two away from the aiming point.

In fact, in these attacks on Skoda the RAF missed the target by some considerable distance. 'I remember particularly a raid on the Skoda Works at Pilsen, duly announced by the

BBC,' grumbled Professor R.V. Jones, head of Britain's wartime scientific intelligence. 'A friendly Czech indignantly told us that everyone in Pilsen knew that there had been no raid, and that the nearest bomb that had fallen was fifty miles away.' During the raid in April the RAF's pathfinders had mistaken a lunatic asylum for the Skoda works, resulting in the main force bombing the open countryside around it.

The experiences of the bomber crews were no easier than those of the fighters, with the main dangers being shot down, fire or the crew being killed or wounded through the fuselage by enemy machine guns, cannons and flak guns. Fortunately, many of Bomber Command's four-engined bombers (such as the Avro Lancaster, Handley Page Halifax and Short Stirling) were extremely robust and could take a lot of punishment before they fell from the sky. The experiences of *Farouk* and her crew were no exception.

On 24 July 1943 the first of six major bombing raids on the German city of Hamburg commenced (comprising four RAF night raids and two USAAF daylight raids). Scheduled for 2/3 August 1943, the appropriately named Operation Gomorrah, the fourth RAF raid, was to complete the destruction of this ancient city. Amongst the 740 RAF bombers was Halifax 'O' of 10 Squadron, No. 4 Group, nicknamed *Farouk*.

Approaching the target, Flying Officer John Jenkins instructed the mid-upper gunner, Aussie Flight Sergeant Arthur Fuller (Royal Australian Air Force), to go to the flare chute on the starboard side ready to release 'window'. This consisted of metallic strips 15in long and 2½in wide in blocks about 4in by 3in deep. When dropped, the bundles scattered into a vast metallic cloud, confusing the German tracking and gunnery radars. The only snag was that it was dangerous for the mid-gunner to be away from his post, especially when approaching a heavily defended target.

Bad weather and electrical storms resulted in only 350 of the 740 planes reaching Hamburg. As Fuller prepared to drop his 'window', *Farouk* was suddenly attacked by a German Junkers 88 night fighter. This versatile aircraft served the Luftwaffe both as a bomber and a day/night fighter and was armed with up to seven 7.92mm machine guns. The Halifax dived to port to escape. Fuller was unable to reach his turret and he was pressed to the floor as the aircraft was hit repeatedly.

The rear gunner, Sergeant Richard Hurst (Royal Canadian Air Force), returned fire, but as the plane levelled out, the Ju 88 returned for a second pass that should have finished off *Farouk*. Jenkins, in an effort to escape, jettisoned his bomb load and dived to starboard as the Halifax was hit again. Fuller managed to crawl to his gun position in time to hear that the German fighter had exploded after being hit by the rear gunner. *Farouk* headed for home as the wind whistled through the extensive holes torn in the fuselage by the German's well-executed attacks. The rear elevators had also been blown away.

Some good did come of the crews' narrow escape. A strong complaint was lodged about 'window' duty, as 10 Squadron was the only operational unit in Bomber Command to use the mid-upper gunner for this operation, so a modification was installed alongside the wireless operator. Jenkins was awarded the Distinguished Flying Cross and Hurst the Distinguished Flying Medal for their bravery.

Hitler was heavily reliant on the oilfields of his allies, notably the Romanian and Hungarian oilfields at Ploesti and Nagykanizsa respectively. American raids ensured output from Ploesti fell and mining of the Danube hampered what was produced being transported to Germany. However, the early raids on Ploesti did not go well. The Americans first attempted to bomb the oilfields in the spring of 1942 but, attacking at night in poor visibility, the bombers failed to locate the target.

The following year, the USAAF conducted Operation Statesman on 1 August 1943 with the intention of bombing Ploesti and the Steaua Rimana refinery at Campina. One of the bomb groups got lost and attacked from the wrong direction, the flak was heavy and the bombers were set upon by German, Italian and Romanian fighters, suffering major losses. This disaster halted USAAF's deep-penetration daylight missions over southern Europe until long-range escort fighters became available in 1944.

Hitler, convinced he needed to centralise his panzer forces, appointed General Heinz Guderian as Inspector General of Armoured Troops in early 1943. Guderian was made responsible for the development, organisation and training of all armoured forces with the Army, Waffen-SS and Luftwaffe, answerable directly to Hitler. To assess the condition of German tank production, Guderian made a point of visiting Daimler-Benz in Berlin and the Alkett Company in Spandau. Their vulnerability to air attack soon became apparent to him.

One of Guderian's first moves was to request that Germany's tank factories be moved before they attracted the attention of the Allies' marauding bombers. At a conference in Munich on 4 May 1943, the Assistant Minister for Armament and War Production, Herr Sauer, opposed this, claiming the enemy was concentrating on destroying the Luftwaffe's aircraft factories. Oddly, Sauer did not consider the tank factories at risk, even in the event of the Allied bombers succeeding in destroying the aircraft plants.

Instead, Guderian had to make do with strengthening the air defences around Cassel, Friedrichshafen and Schweinfurt, Germany's principal tank manufacturing centres. His fears soon came to fruition on 22 October 1943 when the Henschel works at Cassel was bombed and all production came to a temporary halt. Guderian travelled to Cassel to offer his condolences to the workers and their families. The following month, the bombers again turned their attentions on Germany's tank factories, attacking the Berlin works of Alkett, Rheinmetall-Borsig, Wimag and the Deutsche Waffen- und Munitionsfabriken on 26 November.

The US 8th Air Force had flown 1,547 sorties between 17 August and 30 December 1942 for the loss of just thirty-two aircraft, a 2 per cent attrition rate. These sorties were restricted by the limitations of fighter cover. Nonetheless, on 27 January 1943 the USAAF attacked its first German target, Wilhelmshaven, losing only three aircraft. The Focke Wulf assembly plant at Bremen was bombed on 17 April 1943 by a bomber force of 115 aircraft; this time the hot reception cost them sixteen bombers shot down and forty-four damaged.

Then, on 17 August 1943, the USAAF conducted a dual raid on the Messerschmitt aircraft factory at Regensburg and the ball bearing factories at Schweinfurt. This time American losses were even worse. The Germans shot sixty bombers from the sky and damaged another 100, accounting for 370 aircrew. The previous highest number of aircraft lost in a single day had been twenty-six on 13 June over Kiel. The Americans then attacked Bremen, Marienburg, Danzig and Münster and returned to Schweinfurt between 8 and 14 October 1943, losing another 148 bombers and almost 1,500 aircrew. The Schweinfurt raid alone lost 20 per cent of its attacking force.

B-17 bombers attacked Bremen again on 20 December 1943; one specific target was the Focke Wulf fighter plant in the outlying suburbs. It was protected by the full fury of the Germans' air defences. These mounting losses were an appalling wake-up call for the US 8th Air Force. During 1943 it was courage alone that got the bombers to their targets, but guts was simply not enough. Only in 1944, when the long-range P-51B Mustang fighters began to conduct escort duties far over Germany, did the air war over Europe begin to shift firmly in the Allies' favour.

Hitler's factories and his fuel supplies were not the only preoccupation of the Allied bomber fleets. A new German menace had emerged that also delivered death and destruction from the air. Air Vice Marshal Basil Embry, in charge of Bomber Command's No 2 Group, was involved in the destruction of Hitler's so-called V-weapons. Code-named 'Noball', this necessitated bombing the 100 sites being prepared to launch Hitler's rockets and flying bombs against London from November 1943:

Our early attacks on the launching sites from medium altitude showed that although we were able to place our bombs accurately in the target area, it was no easy matter to score a direct hit on a vital part of the site ... In one of our first attacks in the Pas de Calais area we used 46 aircraft without loss, but 14 were damaged by flak. Three days later we attacked the same place with 48 aircraft, and had one shot down and 27 damaged. This was a fair indication of the rapidity with which the Germans strengthened their anti-aircraft defences round a target they thought we were intent on destroying.

German flak steadily increased to counter this bomber effort, as Embry recorded:

Between November 1943, when we began operating against these sites, and mid-May 1944, when the Group was switched to attack targets associated with operation Overlord, we flew 4,710 sorties against No Ball targets, which was the codename given to rocket and flying-bomb sites and their storage and manufacturing centres. In all these sorties we did not lose a single aircraft to enemy fighters, but had 41 shot down and 419 damaged by fire from the ground. Considering how strongly defended many of these targets were by light and heavy flak, and that about a fifth of the missions were flown at minimum height, it was a remarkable record ...

6

SLAUGHTERER AT KHARKOV

On the Eastern Front an equally bloody and protracted air war was fought between the Nazis and the Soviets, but it took time for the Red Air Force to recover from the Luftwaffe's enormously successful surprise attack in 1941. For the Soviet Union there was no equivalent to the battles of Britain or Midway. It was two years later that a largely unnoticed victory over the Kuban began to turn the tide against the Luftwaffe.

Early Soviet fighter aircraft were often found to be completely inadequate against their German counterparts. Similarly, like the Luftwaffe, the Red Air Force only developed a tactical bomber force, so was unable to conduct a strategic air offensive in the same manner as the British and Americans. As a result, while the Red Air Force lacked reach it became particularly adept at ground attack operations employing dive bombers.

By far the best and most famous ground attack weapon in the Red Air Force's armoury was the Ilyushin 'Flying

Tank', the Il-2 Shturmovik, which first appeared in 1941. This was designed as a low-level, close-support aircraft capable of defeating enemy armour and other ground targets. Following early teething problems, it developed into one of the world's most potent ground attack aircraft, armed with cannons, machine guns, rockets and bombs including anti-tank bomblets; with good cause the Germans dubbed it the *Schlächter* or 'Slaughterer'.

The early single-seat Ilyushin Il-2 Shturmovik completed its acceptance trials in March 1941 and 249 had been built just before Hitler's invasion on 22 June. The large undercarriage fairings on the wings left the retracted undercarriage partially exposed, but this meant the pilot could make a belly-up landing without too much harm to the aircraft. Many aircraft were subsequently photographed by the Germans in this state.

This initial model lacked a rear gun position and suffered heavy losses in its opening career against the Luftwaffe's fighters. The two-seat variant first appeared over Stalingrad at the end of October 1942. The following year, the up-engined II-2M was mass-produced and armed with 20mm or 23mm cannon, though the former were faded out. Various upgrades were made to the two-seater; the 37mm gun version had a reduced bombload. This was the Il-2m3 armed with NS-37 37mm cannons, machine guns, bombs and rockets, capable of 251mph (404kmh) and featuring swept-back arrow wings. The Red Army called it the 'Hunchback' and the 'Flying Tank', and it proved deadly against German armour.

Following the Red Air Force's disastrous summer of 1941, General Alexsandr Novikov instigated a major reorganisation in May 1942 that witnessed the creation of independent air armies to replace the corps. These bigger formations consisted of five or more fighter aviation divisions, which offered a far greater punch. Ground support became a primary role alongside air defence. It was in the Caucasus that the Red Air Force began to turn the tide in the air war on the Eastern Front. During April–May 1943, in a series of air battles fought over the Kuban region, the Red Air Force gained local air superiority over the Luftwaffe for the first time.

Notably during the summer of 1943 and the Battle of Kursk, the Shturmovik finally came into its own, severely mauling Hitler's 2nd and 3rd Panzer Divisions. Il-2 pilots perfected 'Circle of Death' tactics for attacking the panzers, usually from behind. They would circle around the enemy armour and peel off to make individual attack runs. When the run ended, they would re-join their formation to wait for another turn. This kept the Germans under constant fire for as long as the Il-2s had ammunition left. It was a truly terrifying experience for the *panzertruppen* battened down inside their tanks. The panzers rocked violently as the Il-2 pilots sought to hit the engine deck, which had the thinnest armour. A direct hit could tear the engine out of the hull.

To support Hitler's attempt to smash the Red Army at Kursk (Operation Citadel) the Luftwaffe massed everything it could spare. General Hans Seidemann mustered 1,000 bombers, fighters, ground attack and anti-tank aircraft in support of the 4th Panzer Army's southern pincer. The northern pincher formed by the 9th Army was allocated another 700 aircraft under Major General Paul Deichman.

In the meantime, the Red Air Force, invigorated by its outstanding success at Stalingrad and growing proficiency following the fighting over the Kuban, felt confident enough to try and pre-empt Hitler's offensive. Just as the Luftwaffe was about to take off from its five airfields around Kharkov on 5 July 1943 it discovered hundreds of enemy bombers bearing down on it. At the very point Hitler's Operation Citadel was launched, the German Army almost lost its vital air support. A force of 132 Shturmoviks, part of a fleet of aircraft, launched a surprise attack on the Luftwaffe's Kharkov airfields.

At the eleventh hour the Luftwaffe was warned by radio monitors, who detected increased communication between the Soviet air regiments. Also, the radar stations at Kharkov reported large formations of enemy aircraft heading their way. These included Shturmoviks and escort fighters from the Soviet 2nd and 17th Air Armies. Panic ensued on the crowded Kharkov airfields and the bombers' departure was hastily postponed. The plan had been that they would take off first and gather over their bases to await their fighter escort.

Instead, German fighters at Mikoyanovka and Kharkov were scrambled and there followed the largest air battle of the war as they intercepted up to 500 Soviet fighters, ground attack aircraft and bombers. General Seidemann reported, 'It was a rare spectacle, everywhere planes were burning and crashing. In no time at all some 120 Soviet aircraft were downed. Our own losses were so small as to represent total victory, for the consequence was complete German air control in the 8th Air Corps sector.'

This was a major setback for the Red Air Force; committing so many fighters to this abortive pre-emptive strike meant that it was unable to challenge Luftwaffe supremacy on the southern flank of the Kursk salient and in the north its response to Luftwaffe attacks was often ineffectual. Certainly in the northern sector, Soviet fighters only began to react to Citadel in the late afternoon and Fw 190s brought down 110 Soviet aircraft by nightfall. Thanks to the poor performance of the two fighter corps with responsibility for providing frontline cover, both had their commanders immediately replaced.

To make matters worse, initial Red Air Force tactics designed to stop Hitler's panzers were flawed. The Shturmoviks failed to get through. The Il-2s and Pe-2s were despatched in small groups lacking fighter escort and these were easily picked off. This was soon remedied using regimental-size formations that were easier to escort and that broke through thanks to weight of numbers. Low-level bombing passes were abandoned in favour of dive bombing at 1,000m at 30- or 40-degree angles. By 8 July, the German Army's advance at Kursk had slowed and the Luftwaffe's control over the battlefield was rapidly declining.

Supported by General Gromov's 1st and General Naumenko's 15th Air Armies, a week later Zhukov launched his counteroffensive with the Western and Bryanks Fronts. These were joined by General Rudenko's 16th Air Army on 15 July when the Central Front went over to the offensive. In just five days the 15th Air Army flew some 4,800 sorties while the 16th managed more than 5,000, over half of which were conducted by Pe-2s and Shturmoviks against retreating German troops.

Alarmingly, by 19 July 1943 Soviet tanks had reached Khotinez, cutting the vital Bryansk–Orel railway. Stukas operating from Karachev supported by other anti-tank planes, bombers and fighters flew to the rescue. For the first time a Soviet armoured breakthrough endangering the rear of two whole armies was driven back from the air. Although during 19–20 July the Luftwaffe's pilots prevented an even larger Stalingrad, this was to be its last major operation on the Eastern Front.

To the south the Red Army then sought to liberate Kharkov and the Red Air Force set about German armour moving up to reinforce their defences. Kharkov was liberated on 23 August and once its airfields were lost the Luftwaffe withdrew to its bases at Dnepropetrovsk, Kremenchug and Mirogorod to help hold the Dnieper Line. It was once again spread over the whole of the Eastern Front providing direct support to the army – for the bomber crews this was a death sentence.

During the last half of 1943, with about 8,500 aircraft, Red Air Force numbers remained static, but over the first six months of 1944 it rapidly expanded to 13,500 planes. The Luftwaffe was slowly but surely overwhelmed. That year the Soviets introduced a new tactical bomber into service, the Tu-2, which was to play a key role in the Red Army's final offensives.

By 1943–44 some 12,000 Il-2s were in service and the Soviets were flying the Il-2m3 variant that included a rear gunner. Similarly, the improved La-5FN and Yak-3 fighters appeared in 1943 and these helped the Red Air Force wrestle air superiority from the Luftwaffe. By the end of the war, a total of 36,183 Il-2 Shturmoviks had been built, more than any other aircraft in history.

7

POINTBLANK

In the run-up to D-Day, Allied fighter-bombers and medium bombers of the British 2nd Tactical Air Force and the US 9th Air Force conducted an intensive offensive against German-controlled railways across northern Europe. Carried out during the spring of 1944, its aim was to hamper Hitler's ability to reinforce his armies in north-west France once Operation Overlord, the invasion of Normandy, was under way.

Controversy has dogged the effectiveness and morality of the Allies' strategic bomber campaign fought over Europe. Winning the war purely through air power was impossible; Adolf Hitler's factories continued to churn out weapons to the very last. The growing shortage of raw materials and oil became a far greater problem, but by then the Allies' ground forces were racing toward the Reich. Only once did the Allies' bomber fleets provide direct and effective support to the ground war and that was through Operation Pointblank – even then the bomber barons chose to interpret the Pointblank directive issued in 1943 largely as they saw fit.

Up until the early months of 1944, the leaders of RAF Bomber Command and the US 8th and 15th Air Forces clung to the hope that smashing Hitler's industries and crushing civilian morale was the way to victory. To the very end, Air Marshal Arthur Harris, commander of Bomber Command, and General Carl Spaatz, commander of the US Strategic Air Forces, remained convinced of their ability to defeat Germany by bombing alone. However, the strategic bomber campaign had not gone well the previous year; British night bombing was notoriously inaccurate (little better than carpet bombing) and ran the gauntlet of German night fighters.

Many crews were thankful that the Lancaster's pre-decessor, the twin-engine Avro Manchester, had been withdrawn from service. This aircraft had proved to be a death trap; its unreliable engines had a nasty habit of catching fire, leading to an alarming loss rate of 40 per cent on operations and 25 per cent on training flights. Likewise the Stirling, which suffered from poor altitude performance (due to a reduction in its wingspan to ensure it would fit inside pre-war hangars), was withdrawn from service in the summer of 1944 and converted to a transport and glider tug.

Yet, the most significant contributions the strategic air forces made in the final phase of the war was the attrition of German fighter defences and the destruction of communications – railways, bridges and canals – in advance of the Allied armies and in the rear of the retreating German armed forces. This transportation plan was thanks to the Deputy Supreme Allied Commander, an airman, Air

Chief Marshal Sir Arthur Tedder, who took up his post on 20 January 1944. It was not until 27 March that he assumed overall direction of the strategic bomber forces and he did not get operational control of the air forces until 14 April. This meant that the strategic and tactical air forces found themselves co-ordinated by Tedder, who was able to prioritise targets with his boss, Allied Supreme Commander General Dwight D. Eisenhower.

Prior to D-Day, it was hard for the Germans to hide their troop and panzer movements along France's roads and railways. In particular, Route Nationale 13 followed the Normandy coastline from Cherbourg in the west to Caen in the east. Before, during and after the Normandy campaign, the Allies' air forces did all they could to interrupt the Germans' lines of communication. In the run-up to Normandy, to prevent the Germans bringing up reinforcements, Rouen's vital rail yards were heavily bombed in April 1944. The city, the historic capital of Normandy, itself suffered extensive damage, causing much ill feeling toward the Allies amongst the civilian population.

Rouen had the dubious accolade of being the first place in Europe to be bombed by the USAAF. American B-17 Flying Fortress bombers initially attacked the city on 17 August 1942, and bizarrely the only casualties were two airmen injured when a pigeon flew into their plane. Rouen was also the recipient of Allied propaganda. Four B-17s showered 800,000 leaflets over Rouen, Lille and Paris in December 1943; this was followed by another 1.2 million leaflets over Rouen, Paris, Caen, Amiens and Ghent, and then 1.3 million on Paris, Lille, Evreux, Rouen and Caen.

Between 1 March and 6 June 1944, marshalling yards in northern France and Belgium were bombed 139 times. The

St Pierre des Corps marshalling yards at Tours were dev-
astated after an attack by RAF Bomber Command on the
night of 10/11 April 1944. Allied light bombers conducted
low-level incendiary raids on German targets, in particular
airfields and communication centres; they also carried out
long-range night-time raids. During 1 May to 5 June 1944
Luftwaffe airfields from the Netherlands to Brittany were
targeted. Similarly, low-level fighter-bomber sweeps were
made over occupied Europe against targets of opportunity.
Over the English Channel, daylight aircraft patrols were
conducted to prevent the movement of light shipping and
coastal convoys.

Attacks on the rail bridges over the Seine and Meuse
that commenced on 7 May 1944 were also designed
to prevent the Germans bringing up reinforcements.
The initial attacks along the Seine included Mantes-
Gassicourt and Oissel, but from the end of the month
onwards ten rail and fourteen road bridges were targeted
as a top priority. By D-Day, from Conflans to Rouen all
the rail bridges across the Seine were down. One failing
of this campaign was not destroying the bridges over
the Loire at Saumar and Tours. Had this been achieved
it would have greatly hampered the 2nd SS Panzer and
17th SS Panzergrenadier Divisions moving north to join
the battle in Normandy.

The Allies deliberately blinded the Germans along the
Channel by knocking out their radars, though this had to
be done in a selective manner so as not to alert the Germans
as to the true location of the amphibious assault. RAF
Typhoon fighter-bombers played a key role in this, striking
sites from Ostend to Cherbourg and the Channel Islands.
To help foster the illusion that the Pas de Calais was the

most likely invasion point, some radars in this area were left alone. Along the coast, out of ninety-two radar sites only eighteen were operational by the time of the invasion. In the three months of April, May and June 1944 the US 8th Air Force also dropped 109,101 tons of bombs in direct tactical support of the Allied armies and on Hitler's V-weapon sites.

Air Marshal Harris threw the full weight of Bomber Command behind D-Day on 6 June 1944:

On the night of the invasion ten batteries in the actual area of the landing had to be attacked, and this took more than 5,000 tons of bombs, by far the greatest weight of bombs dropped by Bomber Command in any single attack up till then. In all 14,000 tons of bombs had to be dropped on the defences of the Atlantic Wall.

On the big day, Air Vice Marshal Embry recalled:

On the morning of D-Day itself, our Mitchell squadrons attacked with great accuracy gun positions directly threatening the approach of the great armada to the Normandy beach-head, and our Bostons flying at sea level laid the smoke screen over the invasion craft.

From now on the task of the Group was to work more closely in unison with the armies which the air forces had helped land ...

Hauptmannt Helmut Ritgen, with the Panzer Lehr Panzer Division, experienced at first-hand the difficulties in trying to reach the enemy's beachheads following the D-Day landings:

> My battalion was attacked by aircraft during a supply halt near Alençon. Bomb and gun bursts set tanks and POL [Petrol, Oil and Lubricant] trucks on fire, soldiers were killed and wounded. Similar incidents happened to all the columns. Some mushroom clouds of smoke were guiding the fighter-bombers to their targets. In spite of increased vehicle distance and dispersion to small groups, marching in daylight under repeated air attack was a risky venture, costing time and losses.

The pilots of the Allied fighter-bombers attempted to wreak havoc on the division. Although there is some dispute as to the exact numbers, losses of more than 200 armoured fighting and wheeled vehicles were reported. While the columns of Panzer Lehr struggled toward their objectives under rolling air attack, its commander, General Fritz Bayerlein, was severely cut up when his car was attacked; his aide and his driver were both killed. He himself got away, slightly wounded but violently shaken.

On D-Day the 2nd Panzer division, under General Heinrich Freiherr von Lüttwitz who had reassumed command on 27 May, was deployed in the Amiens area. It was three crucial days before the division was instructed to move toward Normandy. Because the Allied air forces had destroyed all the bridges over the Seine from Paris to the coast, 2nd Panzer was obliged to make a longer journey when it moved from Amiens to Normandy. Instead of travelling via Rouen, it had to take the detour via Paris, increasing the distance by more than 100 miles. Moving mainly by road and by using the cover of darkness and periods of poor weather, the division managed to cover about 265 miles in two days, an impressive performance in light of the efforts of the Allies' bombers.

A divisional Staff officer with the 17th SS Panzergrenadier Division recalled how moving in daylight would soon draw the unwanted attentions of the Allied fighter-bombers:

> Our motorized columns were coiling along the road towards the invasion beaches. Then something happened that left us in a daze. Spurts of fire flicked along the column and splashes of dust staccatoed the road. Everyone was piling out of the vehicles and scuttling for the neighbouring fields. Several vehicles were already in flames. This attack ceased as suddenly as it had crashed upon us fifteen minutes before ... From now on the 17th SS would travel toward the battle at night, the cost of doing otherwise was simply too great.

Despite Bomber Command's other commitments, principally its attacks on Hitler's weapons factories, raids against his deadly rocket sites were maintained. Air Marshal Harris recorded:

> During the second half of June [1944], Bomber Command dropped more than 16,000 tons of bombs on targets connected with V-weapons, mostly on launching sites, and up to the beginning of September, until the allies occupied the Pas de Calais, a further 44,000 tons of bombs. This 60,000 tons of bombs was equivalent to

one month's bombing at a time when the bomber offensive was at its height.

The heavies, though, proved less effective against the panzers. For example, late in the afternoon of 29 June 1944, elements of the 9th SS Panzer Division were gathering in the woods north of Noyers prior to attacking the British at Cheux, when about 100 Lancaster bombers struck. A huge pall of dust covered the area and it seemed certain that the panzers and panzergrenadiers had been blown to smithereens. However, only about twenty men were killed and by the evening 80 per cent of the armoured vehicles had been dug out and were operational again.

Four American aircraft attacked the main Rouen bridge on 4 July 1944. Two days later, twenty-two American P-47s bombed Rouen and on the 8th six B-17s hit the city's marshalling yards. Subsequent attacks were also planned on Hitler's V-weapon sites in the Rouen area, although poor weather resulted in the mission being called off. On the 18th, German gun positions in the city were dive-bombed; a week later rail traffic to the south of the city was attacked.

Following the German collapse at Falaise in late August, the survivors retreated toward the Seine. The pontoon bridge at Rouen could only take wheeled vehicles and the bridge at Oissel, having been brought down in May thanks to the bombers, was likewise makeshift. Many surviving German tanks and other vehicles that had been coaxed eastward were abandoned on the dockside. On 25 August bombers attacked German transport massed on the quayside twice, the following day the fires were still raging both sides of the river.

One war photographer, on seeing the destruction in the city, remarked:

> If the [Falaise] gap saw the end of the German horse transport, Rouen and the banks of the Seine must have seen the end of an army's motor transport. For about two miles, the banks are crowded with masses of burnt out transport with about 2,000 dead among it.

The Germans were in full flight and their lines of communication were in a state of chaos thanks to Operation Pointblank.

8

TYPHOONS OVER NORMANDY

While bravely attacking a panzer division, Typhoon pilot Flight Lieutenant Godfrey 'Wimpy' Jones, 181 Squadron, ran into intense German flak over Normandy on 16 June 1944, just ten days after D-Day. It proved to be his very last mission of the war. His wife, Betty, was informed that he was missing presumed dead after his plane had come down in a vertical dive from which it never recovered. Such Typhoon losses mounted rapidly during the Normandy campaign. The German Army, lacking Luftwaffe protection as most of their fighters were on the Eastern Front or defending the skies of the Reich, was armed to the teeth with anti-aircraft guns.

In the following months, Mrs Jones, left with a young son, desperately sought to find out what had happened to her husband and for a while clung to the vain hope that he might be missing in action and alive in German hands. What she could not have known was that at the start of the Normandy campaign the RAF had 350 Typhoons;

by the end it had lost 274 aircraft and 151 pilots. This was an appalling attrition rate by anyone's standards. Although more than 3,000 such aircraft were built, with the end of hostilities, those that remained went to the breaker's yard. Yet popular mythology has the Typhoon as one of the key weapons that helped defeat the Nazis following D-Day.

Subsequent analysis of the Normandy battlefield showed that while the gallant Typhoon pilots had daily risked their lives pressing home low-level attacks on the panzers, very few tanks had actually been hit by the aircraft's bombs or rockets. Far from being the much-vaunted 'tank buster' as popular histories suggest, it seems that the Typhoon was a pilot killer. Respected former RAF pilot and noted aviation historian Chaz Bowyer called the aircraft 'brutish', as its remarkable speed and performance came at a deadly price.

The Typhoon was designed as a cantilever low-wing monoplane, as Squadron Leader Hugh 'Cocky' Dundas, Commanding Officer of 56 Squadron, recalled. 'It seemed like an absolutely enormous aeroplane compared with the Spitfire. One sort of climbed up, opened the door and walked in!' Wing Commander Denys Gillam noted, 'I never thought they'd be a good fighter to fighter aeroplane ... but they were a terrific gun platform.'

The RAF's 56 Squadron was first to receive the Typhoon, but due to teething problems the unit did not become operational until May 1942 with an air defence role against low-level intruders. The pilots soon found that the aircraft's poor climb rate and performance at altitude made it little use for anything else. The Fw 190's ascendancy over the Spitfire V meant there was an urgent role for the Typhoon as an interceptor. The squadron found itself acting as test pilots for an aircraft that was still far from combat ready, but this was wartime. There was a sense of newfound camaraderie amongst the pilots, as there always was when they converted to a new aircraft. They knew they were guinea pigs, but inspiration always lay in the name of the thing. Like the Hurricane and Spitfire, the Typhoon was a name to conjure with. It inspired invincibility.

Squadron Leader Charles Green was CO of the second Typhoon squadron, No 266 (Rhodesian), and Sheep Gilroy the third, 609 Squadron, which came under Wing Commander Gillam. This Duxford-based wing was disbanded in September 1942 and the three squadrons despatched to low-level interception duties.

That same month, 181 and 182 Squadrons were issued with the Typhoon 1B. Formed at Duxford on 1 September 1942, following initial defensive operations against German low-level raiders, 181 switched to an offensive role in February 1943. At this time its main targets were Nazi coastal shipping and later ground targets in northern France. Many of its pilots, including Frank Jensen, Leo Arbou, 'Wimpy' Jones and Tom McGovern, had come from 195 Squadron, which had been conducting night intruder interception operations over the North Sea. When the 2nd Tactical Air Force was formed in June 1943, 181 was allocated to it and continued operations as before, but now as part of a mobile tactical wing. From an airfield near New Romney in Kent, they operated over northern France, dive-bombing in particular enemy aerodromes. They were also introduced to rocket firing.

These squadrons soon discovered visibility was problem. More worryingly, the oil cooler failed, leading to overheating, and the engine had a habit of cutting out when coming in to land. Accidents began to occur up to twice a month, with aircraft suddenly dropping out of formation, and at first it was thought monoxide poisoning was to blame. Then the tail started snapping off aircraft descending from high altitudes at speed. The shock usually knocked the pilot out, with fatal results. Again, the tails were coming off when the plane landed heavily. It turned out the balance weight in the rear was coming lose due to vibration. Despite this being fixed, bringing up the tail too quickly on take-off could cause unwelcome swing.

Despite all these setbacks, Squadron Leader Dundas recalled, 'The Typhoon was quite a nice aeroplane to fly, it really was. It wasn't difficult to fly. It was a bit different from the Spitfire to say the least but I couldn't criticise it from a handling point of view.'

Tragically, in the first nine months of its service life far more Typhoons were lost through structural or engine troubles than in combat. Between July and September 1942 it was estimated that at least one Typhoon failed to return from each sortie owing to one or other of its defects. In fact, during the Dieppe raid in August 1942, when the first official mention of the Typhoon was made, they bounced a formation of Fw 190s south of Le Treport, diving out of the sun and damaging three of the German fighters, but two of the Typhoons did not pull out of their dive owing to structural failures in their tail assemblies. The Typhoon's shape also proved a problem, as it was similar to the Fw 190 and this led to some unfortunate incidents with friendly fire. At Dieppe a number of Typhoons were shot down by British flak or fighters mistaking them for Fw 190s.

The Typhoon first used rockets operationally on 25 October 1943 when aircraft from 181 Squadron attacked a target near the city of Caen. The mission was not a great success with three Typhoons lost. Frank Jensen, who had been Wimpy Jones' best man, was shot down and captured, along with Hugh Collins. This was the second time Jensen had escaped a crash; on the first occasion, his Hurricane had been caught by Bf 109s near Manston.

'Wimpy Jones for his part and the three others he was leading,' recalled Tom McGovern, 'carried out successfully the assignment they were given. And all made it back to base, which by then was RAF Merton in Sussex.' Nonetheless, during 1943 low-level attacks resulted in the loss of 380 Typhoons, having shot down 103 German aircraft including 52 Fw 190s. 'In mid-November 181 lost four other of our pilots,' noted McGovern.

Attacks on 'Noball' Hitler's V-1 rocket sites started in January 1944. The following month 181 squadron began to adopt the widespread use of rocket projectiles and carried out operations against various targets in preparation for the forthcoming invasion of Normandy. Roy Crane transferred to 181 Squadron in April 1944 and recalled being involved in cannon and rocket attacks against both V-1 and V-2 sites as well as gun positions, petrol dumps, trains and marshalling yards.

Joining the RAF Volunteer Reserve in 1940, Crane was deferred for nine months before call-up and flying training in the UK and Canada, receiving his wings and commission in December 1942. After operational training on Hurricanes, he joined 182 Typhoon Squadron in August

1943, where sorties included dive-bombing and fighter escort duties.

Tom McGovern remembered, 'In April '44 we launched a variety of attacks over France, which intensified on 22 May, in which we began attacking enemy radar sites on the French coasts prior to the invasion.' In the meantime, Flight Lieutenant Frank Jensen wrote from his German prisoner of war camp on 25 April 1944 to Godfrey and Betty Jones regarding some good news. 'My Dear Wimpy, Just received news of the great event. Please accept my belated congratulations, and best wishes for the future of the Jones' menagerie … My best regards to Betty, and all the boys. Look after yourself old chap.'

Once the rocket attacks had been fine-tuned during May 1944 in the run-up to D-Day, 181 Squadron vigorously pressed home its attacks on the German radar sites. Most notably, nine aircraft from 181, seven from 247 and fourteen from 143 attacked targets near Cherbourg on the 23rd. Once again, 181 suffered casualties; its officer commanding Jimmy Keep was hit and forced to ditch in the sea. He suffered a broken jaw, a broken cheek and bruises, but luckily was rescued by an RAF seaplane.

It was during May 1944 that Kit North-Lewis took command of 181 Squadron. After joining the Army in 1939, he had transferred to the RAF the following year. In August 1941, following pilot training, he was posted to 13 Squadron, flying Blenheim bombers, where he took part in the first 1,000 bomber raids. After a spell with 26 Squadron, flying American P-51 Mustangs, in February 1944 he joined 182 Squadron on Typhoons, as a flight commander. A few months later he was posted to take charge of 181 Squadron. He led his squadron in attacks against the German coastal radar system – including the installation at St Peter Port on Guernsey. He was then to lead the unit into France, where it became part of 124 Typhoon Wing.

To support D-Day, the Commonwealth, European and RAF component of the Allied Expeditionary Air Force was the RAF's 2nd Tactical Air Force. The two largest groups equipped with the Typhoon were Nos 83 and 84, comprising eighteen squadrons organised into seven wings. A third of 83 was made up of Royal Canadian Air Force squadrons and also included 181 along with 182 and 247 Squadrons, which formed 124 Wing based at Hurn in Hampshire.

As well as the Canadians and Rhodesians, there were also Royal Australian Air Force pilots amongst their number. Thomas McGovern DFC (one of just ten Australians to be awarded the French Legion of Honour) flew with 181 from 1943. The Australian official history mentions him as being 'prominent amongst the RAAF pilots engaged in these attacks as well as in missions against railways and airfields'.

During the first five days of June, the Typhoons ensured that all but one of the important coastal radar sites had been put out of action. On D-Day itself, 6 June 1944, the nearest armoured division to the invasion beaches, 21st Panzer, suffered continual air attack and had just seventy tanks left by the end of the day. British forces counted twenty abandoned panzers, with Typhoons claiming another six on the outskirts of Caen. Only six panzers and a handful of infantry made it as far as Lion-sur-Mer to menace the invasion bridgehead. Once the land forces were established in Normandy, the operational task of the Typhoon force was to provide close air support to the British 2nd Army.

With the invasion a success, 181 Squadron, operating from forward airfields in France, was in the forefront of the Typhoons' attacks on the German ground forces in Normandy. On 7 June, North-Lewis led his pilots on three operations to attack tanks and transports. He had a very narrow escape as he flew over Carpiquet airfield, west of Caen, when a bullet shattered his canopy and missed his head by inches. For the next few days his squadron was constantly in action, and in the ten-day period after D-Day North-Lewis flew twenty close support sorties. By the middle of June his squadron was operating from a temporary airstrip near Caen.

However, even in France the Typhoon soon presented further technical problems. Pilots discovered that the large air intake at the front under the fuselage feeding the engine did not like the Normandy dust. This admitted coarse particles and sand into the engine, causing wear in the cylinders and effectively grounding the aircraft. Those Typhoons operating from dirt airstrips, such as 181 Squadron, were soon grounded and the offensive had to rely on support sorties flown from England. The Air Ministry was forced to instruct D. Napier & Sons' Flight Development Establishment at Luton to produce a modification. Napiers responded rapidly to the challenge and general managers Mr Cecil Cowdrey and Mr Bonar designed, built and test flew a momentum-type air filter that was 96 per cent efficient in just ten hours.

Despite the absence of German fighters, both British and Canadian Typhoon pilots were soon experiencing just how fatal German flak could be. The Canadians had lost Flight Lieutenant John Saville of RCAF 439 Squadron the day before D-Day over Guernsey whilst attacking a German radar installation. On 15 June 439 Squadron lost another aircraft, when Flying Officer Jake Ross was caught by flak near Carpiquet. German rounds raked his tail section and the engine caught fire, forcing him to bale out north of Caen.

The following day, Typhoons of 181 Squadron struck enemy vehicles south-west of Tilly-sur-Seulles (west of Caen), held by the Panzer Lehr, which was under attack by the British 50th (Northumbrian) Division. This was no turkey shoot and it was on this occasion that Wimpy Jones' number was up. Panzer Lehr fielded eighteen 88mm flak guns and Panzergrenadier Regiment 901, defending the area, may have been the unit that shot down Flight Lieutenant Godfrey Jones. Also that day, Wing Commander Reg Baker DFC and Bar, OC 146 Wing, came down in Normandy.

Betty Jones was swiftly informed by the Air Ministry on 18 June that her husband was missing in action. 'Regret to inform you that your husband F/Lt GJ Jones 89062 is missing as a result of air operations on 16 June 1944 letter follows stop and stop further information received will be immediately communicated to you.'

Nine agonising days later she was by contacted the Air Casualty Branch confirming he:

Is missing and believed to have lost his life as a result of air operations on 16 June 1944, when he was pilot of one of six Typhoon aircraft which set out on 7.25pm to attack motor-transport and troops south west of

Tilly-sur-Seulles, France. The formation crossed over the target, and during a sudden heavy burst of anti-aircraft fire from the enemy ground defences, your husband's aircraft was damaged, lost height, and dived almost vertically to the ground. Your husband was not seen to bale out, and from the behaviour of the aircraft, it is feared that he had been hit also.

Wimpy Jones was one of many. During the period from D-Day to the end of June, forty-three Typhoon pilots were killed, eight captured, nine evaded capture and seventy aircraft were lost in action. Well over half were lost to ground fire, just 10 per cent were lost to enemy aircraft; only nine enemy aircraft were shot down by them. The following month, thirty-six Typhoon pilots were killed, six were taken PoW and three escaped capture; sixty aircraft were lost in action. During the end of June and early July some Typhoons were diverted from ground attack missions to once again strike the 'Noball' sites in northern France due to increased flying bomb activity.

Then, dramatically, on 17 July Typhoons pounced on Rommel's staff car as it sped along an open road. At a crucial moment in the Normandy battle, Hitler lost one of his key generals. Rommel was hospitalised with serious head injuries and returned home in August (implicated in the 20 July Bomb Plot against Hitler, Rommel poisoned himself on 14 October and was buried with full military honours). The next day, RCAF 439 Squadron lost Flying Officer J. Kalen when his Typhoon exploded as it dived on a target at Mesnil-Frementel. The remains of the plane with Kalen still on board crashed into a nearby forest.

Roy Crane, having completed seventy-one operational sorties, was hit by flak and almost became a victim of the Typhoon's firepower, as he recalled:

Whilst attacking tanks and motorised transport with rockets and cannon in the area of Falaise on 2 August 1944, my aircraft was twice hit by 40mm flak at low level. I baled out and landed in a very hostile Waffen SS camp, lucky to be quickly rescued by two of the nearby German Air Force gun crew that had shot me down. Later that evening I was taken in an open German staff car by the platoon leader from the gun crew, a driver and an armed motorcycle escort in the direction of Falaise. We had only travelled a short distance, when about to pass a column of German tanks, they were attacked by six Typhoons firing rockets and cannon. They came round again and again, leaving terrible carnage. This was an ordeal that has to be experienced to be truly appreciated. They finally got me out of the Falaise pocket to Alencon, after which I was eventually taken after intensive interrogation at Oberursal, to Stalag Luft III at Sagan.

Also that month, Kit North-Lewis was promoted wing commander of 124 Wing, where he remained until the end of the war. Tom McGovern was posted back to England as a tactics instructor to Aston Down, only to be shot down over Germany in March 1945 and captured.

Although Hitler's Normandy counter-attack recaptured Mortain in early August, RAF Typhoons pounced on some 300 armoured vehicles, destroying eight, and other squadrons followed up to take their share of the kills. On 7 August the counter-attack, spearheaded by five panzer divisions, was identified moving against just two US infantry divisions. The panzers had already captured three important villages and were threatening to cut off the US 3rd Army near Mortain as it began moving into Brittany. A shuttle service of Typhoons was established, and by the end of the day they had flown more than 300 sorties, three of them led by North-Lewis. On 8 August at 2115 hours the German 7th Army received orders to postpone the attack following a British breakthrough south of Caen.

Following the German defeat in Normandy, the 2nd Tactical Air Force claimed to have destroyed or damaged 190 tanks and 2,600 vehicles during its sorties over the Falaise battlefield. Typhoon pilot Flight Lieutenant H. Ambrose, 175 Squadron, was amazed by the co-ordination of the air attacks that helped foil German attempts to break out:

[Wing Commander] Charles Green was absolutely brilliant about the Falaise Gap. He had sorted it all out. He saw what was going on and warned the AOC [Air Officer Commanding] and the Army that this was a situation that had to be arrested pretty quickly. Some of the German Army did escape, of course, but the Typhoons and some Spitfires, made mincemeat of the German Army at Falaise. They just blocked roads, stopped them moving and just clobbered them. You could smell Falaise from 6,000 feet in the cockpit. The decomposing corpses of horses and flesh – burning flesh, the carnage was terrible. Falaise was the first heyday of the Typhoon.

During the four months of the Normandy campaign (including the run-up to D-Day), 151 Typhoon pilots were killed, 36 captured and 274 aircraft lost. This was a heavy price to pay, especially when it was revealed that their low-level attacks had not been as devastating as first thought. It has been assessed that only about 100 armoured fighting vehicles were actually knocked out by air strikes during the entire campaign; in stark contrast the Allies lost a total of 1,726 aircraft.

No. 181 Squadron continued in this role, following the Allies through France, into the Low Countries and eventually into Germany itself. It disbanded at Lübeck on 30 September 1945. A year earlier, on 21 September 1944, Tom McGovern and Paddy King made a special visit whilst on leave in England. 'Saw "Wimpy's" widow and kiddie,' Tom recorded in his diary, 'and she was pleased to see somebody who was with him and knew what happened. Little Alan 6 months old is a great little chap. Paddy King and I had tea at their place.'

Betty Jones got the confirmation she had been dreading on 16 February 1945, seven months after her husband had gone missing over Normandy. The Air Ministry wrote he 'was reported to be missing and believed to have lost his life … his death has now been presumed, for official purposes, to have occurred on 16 June 1944.' She never remarried. Her late husband, and all the other Normandy Typhoon pilots killed in action, are fittingly commemorated by the Typhoon memorial at Villers Bocage.

During the four years that the Typhoon was in service, almost 670 pilots were lost (311 of whom were British). By November 1945 3,317 Hawker Typhoons had been built and in total twenty-three squadrons were equipped with the aircraft (twenty-one ground attack and two recon-naissance.) By September 1945 the Hawker Tempest had largely replaced the Typhoons, which were scrapped during 1946–47. The fact that they were scrapped and not sold off as many other aircraft were implies the RAF did not want to pass on the Typhoon's many shortcomings.

9

TARGET TOULON

The 'systematic, especially heavy air attacks on the trans-portation links over the Rhône and Var Rivers,' German Army Group G reported on 7 August 1944, 'point to a landing between these two rivers,' and 'statements from agents confirm this suspicion.' The following day General Wiese, commanding the German 19th Army in southern France, conducted a map exercise at the garrison HQ at Draguignan for all his generals. It was soon clear that the Army was on its own and could expect no help from the Luftwaffe or Navy.

Allied air operations in the south of France consisted of four phases: air ops before D-minus 5; Operation Nutmeg D-minus 5 to 0350 on D-Day; Operation Yokum onwards to H-Hour at 0800; followed by Operation Ducrot. Under Phase I, from 28 April to 10 August 1944, the Allied air forces unloaded 12,500 tons of bombs on the region. Nutmeg began on the 10th and, while concentrating on coastal defences and radar stations, encompassed the whole of the French coast to throw the Germans off the scent.

On the 11th, as the Operation Dragoon assault force began to move from the Naples area toward the south of France, USAAF 12th Air Force B-25 Mitchell and B-26 Marauder twin-engine bombers, and P-47 Thunderbolt fighters, began to strike German targets along the French and Italian coasts west of Genoa. The following day, almost 550 fighter-escorted B-17 Flying Fortresses and B-24 Liberator four-engine bombers attacked targets in France and Italy; the B-24s struck positions in the Genoa, Marseilles, Toulon and Sète areas while the B-17s bombed positions in the Savona area, Italy. Also, more than 100 P-51s strafed radar installations and other coast-watching facilities along the southern French coast. The Germans could hardly miss all this activity.

Two days later, nearly 500 B-17s and B-24s of the 15th Air Force raided positions around Genoa, Toulon and Sète, and struck the bridges at Pont-Saint-Esprit, Avignon, Orange, and Crest, France. In addition, thirty-one P-38 Lightnings dive-bombed Montélimar airfield, while other fighters flew more than 180 sorties in support of the bombers. Also that day, medium bombers blasted coastal defence guns in the Marseilles area while twin-engine A-20 Douglas Bostons, during the night of 12/13 August, attacked targets along the Monaco–Toulon road, and fighter-bombers hit guns and barracks in the area; and fighters strafed airfields at Les Chanoines, Montreal, Avignon, La Jasse, Istres-Le-Tube, Valence and Bergamo.

American medium bombers also hit coastal defences while fighter-bombers pounded various gun positions, tracks, enemy HQ, and targets of opportunity in the

Toulon–Nice area; fighters also strafed radar installations and targets of opportunity along the southern coast as the Dragoon assault forces approached.

Supporting the Allied invasion of the French Riviera in the summer of 1944 were the 42nd Bomb Wing (Medium) and the 17th Bomb Group. The former first saw action during the invasion of Italy, where its units flew close support missions to stop the German counter-attack on the beachhead at Salerno. As the Allied forces progressed, the 42nd took a leading part in attacking Axis road and rail transport, and later in 1944, in the attacks against the monastery at Cassino.

The 17th Bomb Group (comprising the 34th, 37th, 432nd and 9th Squadrons) was involved in the reduction of the Italian islands of Pantelleria and Lampedusa in June 1943; then participated in the invasions of Sicily in July and of Italy in September; and took part in the drive toward Rome. Because of its renowned bombing accuracy, the group was selected to bomb targets in Florence and to avoid the art treasures there. The 17th also took part in the assault on Monte Cassino. In 1943 a heavy bomb group had a total complement of 294 officers and 1,487 enlisted men to fly and support forty-eight heavy bombers; and a medium bomb group had 294 officers and 1,297 enlisted men for sixty-three medium bombers.

The majority of the Luftwaffe units in southern France came under the direction of Fliegerdivision 2. This had its headquarters based at Montfrin, about 18km east-north-east of Nîmes, and as with all Luftwaffe forces in France, was subordinated to Luftflotte 3. Most of the formations had an anti-shipping role. Jagdfliegerführer Süd controlled the fighter defence of southern France. Its HQ had been located at Château La Nerthe (near Châteauneuf-du-Pape, about 10km south-east of Orange) since 1 May 1944.

Nahaufklärungsgruppe 13 (NAG 13), with Fw 190 and Bf 109 fighters, had been deployed in France since 1942, initially at Avignon and then on the Atlantic Coast. In April 1944, it was redeployed to the Riviera, and the 2. Staffel was tasked with maritime reconnaissance between the Spanish border and Corsica. By mid-1944, NAG 13 was equipped with a mixed unit of Fw 190 A-3/U4s and Bf 109 G-8s, their armament reduced to two machine guns and a single cannon respectively. The anti-shipping Stab and III/ Kampfgeschwader 100 by 1944 was based on the airfields of Blagnac and Francazal, both near Toulouse.

Luftflotte 3 issued orders on 8 August that, in the event of an Allied landing in southern France, local air units would be placed under Fliegerdivision 2's control. In the face of a landing, Luftflotte 3/Fliegerdivision 2 planned to bomb the invasion fleet at first light with all available aircraft. However, the Allies had mastery of the air and all regional commander General Blaskowitz could do was stand back and watch. On 25 June he may have witnessed 250 USAAF B-17 bombers attack the airfields near Toulouse with impunity.

The key naval port facilities at Toulon were second only to those at its larger western civilian neighbour at Marseilles. Its defences against attack from the sea included batteries at Mauvannes and on the peninsula of Saint-Mandrier. Inland, Toulon is partly shielded by the hilly terrain and narrow river valleys running west of the port from Bandol to the Grand Cap Massif and the region west of Solliés-Ville. The eastern approaches running from La Valette to La Crau and Le Pradet are more vulnerable because of the coastal plain that provides the best route of attack.

Before the German takeover of southern France in November 1942, on the French side, as a token of goodwill towards the Germans, coastal defences were strengthened to safeguard Toulon from an attack from the sea by the Allies. These preparations included plans for scuttling the fleet in case of a successful landing by the Allies (most of the French fleet was scuttled when the Germans took over).

On the peninsula in front of the harbour at Toulon was a complex of three massive 340mm gun batteries mounted in turrets. In addition, numerous medium-sized guns were strung out along the coast, totalling seventy-five flak guns. Two 340mm guns from the scuttled battleship *Provence* had been moved by the Germans to the Cape Cépet battery on the Saint-Mandrier Peninsula, which forms Toulon's large bay. Originally there had been two twin gun turrets in the battery, but when the French fleet was scuttled they had been damaged. The Germans managed to repair two of the guns, though one was sabotaged just before the landings.

The 17th Bomb Group attacked the Toulon harbour gun complex twice on 13 August, encountering intense, accurate anti-aircraft fire that damaged a number of the attacking B-26 Marauders. The heavy Allied bombing of Toulon and other targets in the days before the landing alerted Army Group G's commander General Johannes Blaskowitz that something was likely to happen in this area. Indeed, by the 14th, suspecting an attack in the Marseilles–Toulon region, Blaskowitz moved the 11th Panzer Division and two infantry divisions east of the Rhône just in case.

On 12 August, due south of Ajaccio, Corsica, the Luftwaffe picked up two large convoys, each of about

seventy-five to 100 merchant vessels and warships including two aircraft carriers, heading north-east to the harbour;
already in the harbour were another twenty vessels.
Confirming that an invasion build-up was taking place, on
the airfield were sighted eight gliders and five multi-engine
aircraft. Luftflotte 3 ordered that reconnaissance efforts
over these convoys be stepped up day and night.

Two days later, when an Fw 190 fighter and four Bf 109s
were on convoy patrol south of Marseilles-Toulon-Golfe
du Lion, no sightings were made. However, subsequently
at 1915, the Luftwaffe reported landing craft stretching
50 miles west from Ajaccio Roads and at 2035 two convoys
were sighted 100 miles south of Menton, numbering more
than 100 landing craft as well as surface and air escorts.
Invasion was imminent.

In the meantime, twelve P-38s of the 94th Fighter
Squadron, 1st Fighter Group dive-bombed the HQ of
Jagdfliegerführer Süd at La Nerthe. At 1900 hours the base
reported that its command post had been destroyed and
that three personnel were killed, three badly wounded
and three lightly injured. The HQ was rendered inoperable
and with the phone lines out, the base commander decided
to set up an aircraft reporting centre in Courthézon (6 miles
south-east of Orange) the following day.

At noon on 13 August, the invasion convoy sailed
from Naples through the Sardinia–Corsica Straits and
deployed offshore of the Riviera beaches at dawn on the
15th. The naval guns and bombers bombarded the coastline as the landing craft were lowered and the first waves
of troops were ferried toward the assault beaches. Early
on 15 August, in the 15th Air Force's first mass night raid,
252 B-17 Flying Fortresses and B-24 Liberators, after a

pre-dawn take-off, bombed the beaches in the Cannes–Toulon area in immediate advance of Operation Dragoon. Twenty-eight other fighter-escorted B-17s attacked highway bridges over the Rhône River, however other B-17s sent against coastal gun positions had to abort their mission owing to poor visibility. In the meantime, 166 P-51 fighters escorted transport aircraft carrying the airborne invasion troops.

With the weather remaining generally good, carrier-based planes were able to conduct regular spotting missions and attacked inshore targets, including gun emplacements and railway facilities, with impunity. Over the following week, the carrier USS *Tulagi*'s aircraft flew a total of sixty-eight missions and 276 sorties, inflicting considerable damage on the enemy.

One squadron from the *Tulagi* alone reported a record of 487 motor vehicles ranging from staff cars to panzers destroyed and another 114 damaged. In the first week of Dragoon, General Blaskowitz lost 1,500 vehicles destroyed, mainly to air attack, 200 vehicles captured and 1,500 horses killed. Losses in men comprised 1,000 dead and 3,000 captured. During the fighting of 15–18 August, the remnants of the 242nd Infantry withdrew into Toulon.

Brigadier General Saville was full of praise for the air support effort from the carriers and wrote to Vice Admiral Hewitt to show his gratitude:

> I would like to express my appreciation for the outstanding work they have done and for their perfect co-operation. I consider the relationship and co-operation of this force to be a model of perfection and a severe standard for future operations. Today, I

personally counted 202 destroyed enemy vehicles from four miles west of Saint-Maximin to two miles east of LeDuc. Well done and thanks.

Phase III/IV of Operation Dragoon's air war commenced on the 16th. The USAAF's 17th Bomb Group was called on to destroy Toulon's heavy guns. Understandably, the wing was sceptical of its ability to score direct hits on such small targets in the face of concentrated anti-aircraft fire. Pilots recalled it as one of the toughest targets of the war. The group was also tasked to destroy a number of bridges over the Rhône and Durance Rivers. It was to lose five aircraft shot down and numerous others damaged during this phase of the air campaign.

The 37th Bomb Squadron under Captain Rodney S. Wright, from Washington State and formerly an RAF pilot, was the lead formation attacking the flak guns and emplacements around Toulon. His number two on his right wing was Maurice Walton, piloting Red 34, a B-26 Marauder. The squadron was making its final approach when suddenly aluminium chaff (designed to distract enemy radar) began to flutter through the formation. This simply helped to highlight the exposed bombers as they made their run into the already fierce flak.

Red 34 had just released its bombs when the right engine was hit and caught fire. Luckily, co-pilot Don Hoover was able to feather it and this action combined with a fire extinguisher did the trick. Unfortunately, tail gunner Sergeant Jesse A. Ward was hit and his right arm torn open. When he did not respond to the crew check, Staff Sergeant Brown, the waist gunner, went to assess Ward's condition and found him bleeding profusely. Bombardier

Tom Richardson then crawled back and gave him an injection of morphine. Staff Sergeant Chuck Zahn, top turret gunner, helped move Ward to the radio compartment, where Richardson administered first aid.

The stricken bomber was forced to fly for two hours on one engine and Sergeants Brown and Zahn, in a desperate effort to lighten the load, threw out guns, flak vests, ammunition and anything that was excess weight. All the while, Don Hoover held the aircraft on course at about 170mph over the Mediterranean. Red 34 had suffered so much damage that it belly landed upon return to base; luckily it did not nose over as happened to some of the other crews that had bellied up. Tail gunner Sergeant Ward, who survived the war, gained the Purple Heart and Distinguished Flying Cross.

Sergeant Delbert F. Kretschmar, with 95th Bomber Squadron, 17th Bomb Group, reported things differently on 16 August: 'Mission was guns at South of Toulon. France. Mission was OK. Little flak and was inaccurate.' However, on 18 August another B-26 went down and two days later the 37th, 95th and 432nd each lost an aircraft.

A-20 bombers hit motor transport in the Nice area during the night of 21/22 August and hit industrial buildings in southern France during the day. Fighters also attacked motor transport west of the Rhône and in scattered parts of south-east France; the 85th, 86th and 87th Fighter Squadrons, 79th Fighter Group, were deployed from Corsica to southern France with P-47s; and the 315th Fighter Squadron, 324th Fighter Group, moved from Corsica to Le Luc, also with P-47s.

On the ground in the eastern sector of Toulon's defences, Free French troops attempted to capture La Poudrière, where the defenders employed old French tanks. On the 23rd attacks were stepped up on the Touar ridge, which was secured. By the evening the French were in force outside and inside the city. They took great pleasure in raising the tricolour over the sous-Préfecture building. At 2245 Admiral Rufus the city commandant agreed to surrender Toulon unconditionally.

The 'Battle of Toulon Harbour' cost the 42nd Bomb Wing eight B-26s and resulted in battle damage to 125 other aircraft. During these missions, the 17th Bomb Group encountered the heaviest, most accurate, flak it had ever seen. Of the twenty-eight raids that the 42nd Wing conducted against this complex, only five succeeded in making a dent in the troublesome batteries. It was not a good result for the 'Bomber Barons'.

10

DEATH OVER THE REICH

Flight Sergeant Norman Jackson, 106 Squadron RAF, was involved in one of the raids on Schweinfurt and only just survived:

I was shot down on the 26th April 1944. Our target was the ball-bearing factory at Schweinfurt. We'd had a lot of goes at that already. Anyway, we'd got our bombs down, but the flak was coming up and there were fighters all around … I saw flames coming from the starboard inner engine, so I grabbed the fire extinguisher … I got out and slid down the wing … Then I was shot off the bloody wing, and they threw the parachute out of the plane.

For his bravery Jackson was later awarded the Victoria Cross. These mounting attacks forced Hitler to intensify his air defences.

When Speer presented his situation report to Hitler at the end of June 1944 it was not good news. If Germany's synthetic oil production plants and the Hungarian and Romania oilfields were not adequately protected from the Allies' unrelenting bombing, then by September everything would grind to a halt with tragic results. In response, Hitler ordered the flak and smoke defences be increased, but the real problem was the Luftwaffe's inability to prevent the Allied bombers pressing home their attacks. Even though Speer had ensured German fighter production was up, it could not keep pace with the heavy losses and in part that was due to the Allies bombing the aircraft factories and oil refineries.

To defend the skies over the Reich, around 75,000 Luftwaffenhelfer (schoolboys) from the age of 16 were drafted into flak school to help man flak batteries. Also, 15,000 girls and women, along with 45,000 Russian prisoners of war and 12,000 Croatian soldiers, were likewise conscripted into air defence duties. Seventeen-year-old Friedrich Kowalke served in the Luftwaffenhelfer defending the Magdeburg area with Lemsdorf battery deployed to the north:

When the first bomber formations were approaching the Münster-Osnabrück area the civil population was warned. Correspondingly to that penetration the command of 'Luftflotte Reich' normally gave the order that aircraft of fighter Division 1 had to assemble over Magdeburg …

Sometimes we saw dense condensation trails or heard the continuous deep roaring of bomber formations in the distance, interrupted by the high singing noise of escort fighters. Also, P-51s were seen hunting German fighters (Me 109s) on the deck.

Despite constant and intense bombing throughout the second half of 1944, Speer managed to keep weapons production on an even keel. The only decline was in tanks. During June, July and August his factories constructed 2,430 panzers; over the next three months this slipped to 1,764 and almost 400 of these were not delivered due to the disruption of Europe's rail networks. USAAF attacks on the Henschel works at Cassel deprived the Germans of at least 200 Tiger tanks.

This decline in production was partly offset by the increase in assault gun output. These factories were mainly located in Czechoslovakia, where they remained largely immune to bomber attack, and assault gun production rose from 766 in August 1944 to 1,199 by November 1944. The first real threat to the vital Czech tank plants did not occur until August 1944, when there was a rebellion against the pro-Nazi Slovak government. For Hitler this was a potential disaster as it was not only a threat to the Skoda and BMM tank plants in neighbouring Bohemia and Moravia, but could also impede the retreat of his forces. Troops were sent and the rebellion swiftly crushed.

On 5 September 1944, Speer reported again and things had not got any better. Germany's weapon factories were running out of energy supplies and raw materials. Based on the assumption that Hitler was prepared to abandon the occupied territories in northern and southern Europe, Speer's industries could continue production for about a year before they reached crisis point.

Hitler refused to accept the situation and refused to make any withdrawals; he knew to do so would mean an increase in air attacks on Germany's heartland. Speer, realising that the Third Reich was doomed, began secretly to make what preparations he could to save Germany's industries from the Allies and Hitler.

The Allies' strategic bomber campaign really began to take its toll on Italy's factories and infrastructure after the German occupation of northern Italy in the summer of 1943. Continuous Allied air attack made it almost impossible for the Germans to receive new motor vehicles and adequate supplies of fuel from Germany. In addition, the production of motor vehicles by Italian factories was seriously impeded. Prisoners captured by the Allies in Italy in 1944 reported 'that German trucks were habitually overloaded, and when they broke down could with difficulty be repaired because of the shortage of spare parts created by your strategic bombing of the factories in Milan and Turin'.

Intense Allied bombing of the Italian rail network forced the Germans to rely increasingly on trucks to move their supplies from the Florence area. In the first three weeks of April, an American fighter group claimed to have destroyed or damaged more than 400 German trucks caught on Italy's open roads. The official British history of the campaign records:

> The destruction of motor-vehicles was so enormous and the shortage of petrol so severe, the German Armies were compelled to rely increasingly on horses and oxen to move their transport, and to commandeer farm wagons, urban buses, and civilian cars of every description.

In early 1944 General Spaatz, commanding the US strategic air forces, proposed concentrating on Hitler's fuel supplies; this though competed with the priorities of preparing for D-Day and destroying the V-weapon sites. Romania's Ploesti oilfields came under sustained attack in mid-April that year and within six weeks had been raided twenty times. By mid-1944 Germany was receiving just 10 per cent of the 2.5 million tons it had been getting the year before from Romania and Hungary. Imports plummeted from 200,000 tons in February 1944 to just 11,000 tons by the summer. When the Red Army swept into Ploesti in late August 1944, the Germans failed to destroy the oilfields. Their loss mattered little as by this stage they were contributing nothing to Germany's industries and the Wehrmacht.

The Allies also conducted a concerted air campaign to smash Hitler's synthetic oil production. The aim was to keep the Luftwaffe from the skies and cripple the Wehrmacht's panzers. They bombed the synthetic oil plants at Brux, Bohlen, Leuna, Lutzendorf and Zwickau on 12 May 1944. After two years of largely indifferent strategic bombing, the results of these missions were immediate and spectacular.

Aviation fuel production for the Luftwaffe was brought to an almost complete standstill. The Germans had produced 156,000 tons in May 1944; in June it had dropped by two-thirds to just 52,000 tons. The RAF and USAAF kept up the relentless pressure and in July aviation fuel production fell to 35,000 tons and in August to 17,000. By January 1945 the figure stood at 11,000 tons and production had ceased by March. By September 1944 the Luftwaffe was left with just five weeks of fuel and Germany's arms industries had soaked up the last of the raw materials, which could not be replaced. The Third Reich was haemorrhaging to death and there was nothing Speer could do about it.

The production of gasoline for vehicles plummeted from 134,000 tons in March 1944 to just 39,000 tons in March 1945. This systematic and concerted destruction of the synthetic oil plants meant that by September 1944 German petrol stocks, which had stood at 1 million tons in April, had fallen to 327,000 tons. Similarly, the production of diesel oil fell from 100,000 tons to 39,000 tons over the same period. The German industrial heartland of the Ruhr was also pulverised.

A Halifax bomber of 425 Squadron, Royal Canadian Air Force, was not as lucky as *Farouk*. At 8.30 p.m. on Saturday, 4 November 1944, the aircraft, nicknamed *Smitty's Kite*, was only 6 miles from the town centre of Bochum along with 700 other Halifaxes and Lancasters. The crew consisted of Flying Officers Donald Smith (pilot), Ermine Knorr (bomb aimer), 'Jamie' Jamieson (navigator) and Flight Sergeants 'Wally' Clowes (flight engineer), Jim Gale (tail gunner), Al Limacher (mid-upper gunner) and Bob Ford (wireless operator).

The German manufacturing town of Bochum was in the Ruhr and had a population of about 300,000, many of whom worked in the surrounding coal, iron and steel industries. The raid was scheduled to last thirty minutes and the ground defences were to be swamped. *Smitty's Kite* was lumbering along at almost 300mph with 11,000lb of bombs, including one massive 2,000-pounder and six 1,000-pounders, as well as canisters of 4lb incendiaries; the fuel tanks held more than 2,000 gallons of high-octane aviation fuel. Fortunately, there was only light flak.

It was always a dangerous period for the bomber crews when they had to hold a steady course at a constant speed and height for two or three minutes whilst over the target. It was then that *Smitty's Kite* came to grief. A sudden explosion under the inner starboard engine caused a sheet of flame to be blasted into the fuselage. Shrapnel ripped open the starboard side of the bomber, fatally striking flight engineer Clowes, who happened to be standing next to the pilot. This saved Smith's life and prevented an even greater disaster.

The automatic fire extinguishers did not cut in and when Smith tried to cut the fuel, this failed, leaving the starboard wing brightly ablaze, illuminating the stricken bomber for the German searchlights and flak guns. Smith gave the bale out signal. Jamieson and Ford escaped via the lower floor hatch, as did Knorr. Gale successfully baled out from his rear gun turret. Limacher in the mid turret was stunned, but the flames quickly roused him and he dropped down into the body of the plane.

Smith stayed at the controls, but he was unable to stop the aircraft going into a spin and the bomber exploded. Limacher, in the process of exiting, found his feet were trapped in the debris. Leaning from the hatch, he pulled his chute, which luckily yanked him clear. Smith was blown through the roof of the cockpit, which prevented him from hitting the propellers.

As he tumbled into space, Smith feared he had no parachute. His right arm was broken and he could not reach the ripcord with his left. He hurtled 12,000ft; managing to pull his life-saving cord at the last minute, he crashed into a cabbage field, breaking a rib. All of the crew, except for the unfortunate Clowes, went into captivity. Some twenty-five bombers were lost that night.

Flight Lieutenant Joe Herman, of 466 Squadron, Royal Australian Air Force, also taking part in this huge raid, had a miraculous escape that was a sheer chance in a million. His Halifax bomber was hit by the 'light' flak and caught fire. Then, to make matters worse, two more hits took off the outer sections of the starboard wing and the plane went into a terminal spin.

At 17,000ft the aircraft blew up and Herman, who had not clipped on his parachute, plummeted to certain death. He fell 12,000ft in complete free fall, when suddenly as if by divine intervention, he collided with John Vivash, his mid-upper gunner, whose chute had opened. Grasping this chance to save himself, Herman frantically locked his arms around Vivash's legs. When the pair hit the ground, Herman escaped with two broken ribs.

When the time came, Field Marshal Model commanding Army Group B, charged with defending the Ruhr, with the support of Albert Speer, refused to destroy Germany's remaining bombed and battered industries. By that stage the factories in Czechoslovakia and France had long been liberated by the Allies. Clearly the strategic bomber campaign's greatest contribution to the defeat of Hitler was the destruction of the fuel stocks that fed his armies and factories.

Had the 'Bomber Barons' attacked Hitler's oil supplies earlier it would undoubtedly have shortened the war. Speer supported this after Germany's capitulation, stating:

The Allied air attacks remained without decisive success until early 1944. This failure, which is reflected in the armament output figures for 1943 and 1944, is to be attributed principally to the tenacious efforts of the German workers and factory managers and also to the haphazard and too scattered form of attacks of the enemy who, until the attacks on the synthetic oil plants, based his raids on no clearly recognisable economic planning ... The Americans' attacks, which followed a definite system of assault on industrial targets, were by far the most dangerous. It was in fact these attacks which caused the breakdown of the German armaments industry.

The air battle for the Reich had been won at a terrible cost. Between 1939 and 1945 Bomber Command lost 52,000 dead, 20,000 wounded and 11,000 captured as well as 7,172 aircraft. Another 8,000 men and women were killed in accidents. The USAAF lost 121,867 personnel, of whom 40,061 were killed, 18,238 wounded and 63,568 captured or missing. Some 41,575 aircraft were lost in combat and due to accidents. German day and night fighters were credited with the destruction of 70,000 Allied aircraft on all fronts, of which some 45,000 were on the Eastern Front. For those airmen who did survive the air war, it was largely down to luck as much as professionalism.

At the end of the war Air Marshal Harris was quick to justify his methods. He measured his success by the thousands of acres of devastation that had been inflicted on Germany's cities. He chose to question the veracity of the USAAF's tactics and the results. Harris clung to the belief that area bombing produced the desired results rather than singling out specific targets. He also argued that the bomber offensive had been hamstrung by other priorities:

Moreover it was only in the last few months of 1944, just when production in Germany began to fall most rapidly, that we were allowed to use any considerable part of our force against German industrial cities. Over the entire period of the war only 45 per cent of the Command's whole effort was against German cities, so that in fact we were using for the main offensive a force which was not only less than one-quarter of the strength originally planned, but nearer one eighth.

He felt, had they had the same force in 1943 along with American bombers and without interference, then 'Germany would have been defeated outright by bombing as Japan was …'

II

MIG ALLEY

Jet-powered aircraft first appeared in the closing stages of the Second World War with the German Messerschmitt 262 and the British Gloster Meteor. They offered fighter pilots greatly enhanced speed. The twin-engine 262 did not see combat until October 1944 and its engines soon developed a reputation for being unreliable. These were also inhibited by just twenty-five hours of service life. The aircraft's development had been hampered by Hitler's insistence that it be employed as a bomber instead of a fighter. While it ended up being used in both roles, its greatest asset was its 540mph (870kmh) speed, enabling it to outrun all Allied fighters.

About a dozen Meteors, also with twin engines, became operational in the summer of 1944, flying anti-V-1 patrols. The improved Mk3 was not delivered until the end of the year and flew ground attack missions in the closing weeks of the war. It was powered by Rolls-Royce Derwent turbojets. Whereas the 262 fought against the Allies' air forces, the Meteor never went into action against the Luftwaffe. Post-war versions were faster than the 262 and stayed in service with the RAF well into the 1950s.

In the closing stages of the Chinese Civil War, the communists were supported in their struggle against the nationalists by the Soviet Union's very first effective jet fighter – the MiG-15. In the skies above Shanghai, Soviet MiG-15s fought to fend off nationalist bombing raids launched from their sanctuary in Taiwan. Captain Kalinikov, of the Soviet 50th Fighter Aviation Division, shot down a P-38 belonging to the nationalist air force on 28 April 1950. This was the first aerial victory for the MiG-15. On 11 May, Captain Schinkarenko claimed a B-24. Shortly after, MiG-15s took part in the very first jet-versus-jet dogfights during the bloody Korean War.

The MiG-15 jet fighter came about through one of those strange quirks of fate. Britain, on the very eve of the Cold War, handed over the Rolls-Royce engines with which to power it. Moscow, drawing on its experience with the highly flawed MiG-9, first flew the vastly improved MiG-15 in late 1948, and it was accepted into service the following year. It was designated 'Fagot' by NATO, maintaining the 'F' prefixed code names started with the MiG-9 'Fargo'.

This aircraft heralded a remarkably successful trend with subsequent Soviet MiG fighters and Sukhoi ground attack aircraft. It was armed with both 23mm and 37mm cannons in order to counter America's B-29 Superfortress strategic bomber. Moscow was able to conduct valuable air-to-air combat trials using the Tu-4 Bull, a Soviet-cloned B-29.

The MiG-15 was quickly sold to eager customers, with China receiving an improved version in 1950. These cut their teeth during the Korean War, where the MiG-15's capabilities came as an unpleasant wake-up call for the West and highlighted the threat posed by rapidly developing Soviet fighter technology. Western Intelligence was equally alarmed when it was confirmed that Soviet-piloted MiG-15s were indeed shooting down B-29s over Korea.

After the Second World War the Soviet Union initially struggled to develop an adequate jet fighter because it was reliant on captured German technology. The net result was that Mikoyan-Gurevich's MiG-9 was woefully underpowered and difficult to control. It used a pair of reverse-engineered German BMW jet engines that simply did not generate enough thrust.

In response to this problem, Soviet Aviation Minister Mikhail Khrunichev and aircraft designer A.A. Yakovlev suggested to Stalin that they get their hands on the more powerful British Rolls-Royce's Nene jet engine. This could then be copied. And so the British Labour government of the day decided it was quite happy to supply a licence and all the technical data that the Soviets needed. Few in Britain seemed to heed the likely outcome of such generosity.

Very quickly a Nene clone known as the Klimov RD-45 went into mass production. Later Rolls-Royce demanded more than £200 million in unpaid licence fees; Moscow ignored the request. It was decided to build two different MiG-15 prototypes: one was based on the MiG-9, which featured the traditional straight wings; the other had swept wings and a tailpipe that ran back to the swept tail. The Soviets had captured examples of the German Me 262, which was the first fighter with an 18.5-degree wing sweep, and they had also laid their hands on the proposed plans for the Focke Wulf Ta 183, which had never been built.

The MiG-15, with a mid-mounted 35-degree swept wing, had greater similarity to the proposed Ta 183 than the contemporary American F-86 Sabre jet fighter. The prototype was designated the I-310 and first took to the air in 1947, just two months after the F-86. Once green-lighted and designated the MiG-15, the first production model took to the skies on 31 December 1948.

An improved variant, the MiG-15bis (i.e. second), entered service in early 1950. This carried cannons and could take unguided rockets and bombs. Although well powered, the tail design hampered control as the aircraft approached Mach 1. In addition, it had a nasty habit of spinning after an engine stall, from which the pilot could not recover. Another alarming glitch was that once half empty the fuel tank could develop negative pressure and explode. The MiG-15's original task was to catch and destroy American bombers, and to achieve this the twin 23mm guns each had eighty rounds and the single 37mm had forty.

In response to the American-led United Nations inter-vention in the Korean War, the Soviet Union quietly despatched its 64th Fighter Aviation Corps, equipped with the MiG-15, to patrol the Chinese–North Korean border along the Yalu river. Stretching from the Korean cities of Sinuiju in the south-west and Kanggype in the north-east, this area became known to UN pilots as 'MiG Alley'. The MiGs would pounce on American B-29 bombers targeting North Korea's military industries. Moscow also agreed to provide the Chinese and North Korean air forces with the MiG-15 as well as pilot training.

The Soviet MiG-15 jet fighter first appeared over Korea in late 1950, with initially some fifty aircraft flown by Chinese and Soviet pilots in support of the North Koreans. On 1 November 1950, eight MiG-15s intercepted fifteen United States Air Force F-51D Mustangs. One of the Mustangs was shot down. Then, in the first jet-versus-jet victory, the same unit shot down an American F-80 Shooting Star. To counter the presence of the MiG-15, three squadrons of the F-86 Sabre, America's only operational jet with swept wings, were sent to Korea.

From late 1950 the communists made strenuous attempts to gain control of the air from the more experienced UN pilots. The futility of this soon showed in their casualties. For the loss of 114 planes, the UN destroyed or damaged 2,136 communist aircraft, including 838 MiG-15s – there were another 177 probables. During the closing stages of the war, thirteen MiGs were being shot down for every American F-86.

Nonetheless, the MiG-15 threat did put an end to daylight bombing raids, which were replaced by night radar-guided operations. By 1951 there were two regiments of MiG-15bis operating as night fighters over Chinese airspace, which were used to intercept USAF reconnaissance flights. In the first five months of 1951, the F-86 Sabre 4th Wing flew 3,550 sorties and claimed twenty-two victories; not a single Sabre was shot down by MiGs, though a number were lost to accidents.

During June 1951 the communists grew more confident and, in some fierce engagements, succeeded in downing several Sabres. The explanation for the increased effectiveness of the MiG fighter units was the arrival in Manchuria from March 1951 onwards of Czech, Polish and Russian pilots on three-month combat tours. At this stage Soviet Air Force MiGs were repainted in Chinese People's Air Force markings, but later MiGs openly displayed the plain red star insignia of the Soviet Union.

Enough MiG-15s were gathered in the Yalu area by September 1951 for the Soviets and Chinese to deploy the aircraft outside their Chinese sanctuaries. On the 25th of that month, sixteen Chinese-piloted MiG-15s engaged US Sabres. North Korean pilots got in on the act the following year.

The Soviet 64th Fighter Aviation Corps sent secretly to fight for North Korea in November 1950 consisted largely of veterans or aces from the Second World War. They claimed more than 1,300 UN aircraft while losing 345 of their own. Amongst these Soviet pilots were sixteen aces, the top scorer being Evgeni Pepelyaev with twenty-three kills. His closest rival was Captain Nikolay Sutyagin.

Sutyagin fought as a fighter pilot during both the Second World War and the Korean War. He was one of the top fighter aces fighting alongside the Chinese and North Korean air forces. He reportedly flew 149 combat missions, resulting in sixty-six aerial engagements during which he personally shot down a total of twenty-one enemy aircraft (plus two others shot down in a group). These victories consisted of fifteen F-86 Sabres, two F-84 Thunderjets, two P-80 Shooting Stars and two Australian Gloster Meteors. Sutyagin was awarded the Order of Lenin, three Orders of the Red Banner and

Order of the Patriotic War First Class, and made a Hero of the Soviet Union.

Despite the MiG-15's presence, the American F-86 Sabre achieved almost indisputable dominance over the skies of Korea, North and South. Sabre pilot Captain Joseph McConnell became a fighter ace with sixteen kills. UN forces, though, did not have it all their own way. Operation Strangle, designed to cut North Korean supply lines, cost the UN 343 aircraft lost and 290 damaged.

Overall, communist pilots had few successes, however, as their gunnery was often poor, and in dogfights MiGs were not infrequently seen to spin out and crash. This was probably because the MiG had no provision for G-suits and was rather unstable, giving no warning of an impending stall, and had a slow roll rate. By the time of the ceasefire in July 1953, there were 297 Sabres in Korea facing around 950 Sino-Korean MiGs. During the conflict, F-86 pilots claimed to have destroyed 792 MiGs in air-to-air combat for the loss of seventy-eight Sabres – a phenomenal ten to one kill-to-loss ratio.

Crucially the North Korean, Chinese and Soviet air forces had little bearing on the ground war; rather they spent much of their time engaged in aerial dogfights. Fighting enemy jets was one thing, but pressing home attacks on well-defended ground forces often with numerous anti-aircraft guns was another matter. Throughout the war, the North Korean Army lacked fighter cover as well as fighter-bomber and bomber support.

Remarkably, after the Korean War the MiG-15 became one of the most widely built jet aircraft ever produced.

More than 12,000 were built in the Soviet Union, with around another 6,000 manufactured under licence. Egyptian Air Force MiG-15bis and MiG-17s saw action during the Suez Crisis in 1956. It was also used as a jet trainer during the Vietnam War.

12

DAY OF THE HELICOPTER

The Germans had employed very limited numbers of early helicopters for military purposes in the Second World War. Likewise, the Allies made restricted use of the Sikorsky R-4 helicopter. However, it was not until the Korean War that the helicopter began to show its true potential thanks to its vertical take-off and landing capabilities. It was to come of age in Vietnam with the development of air mobile warfare. Thanks to innumerable Hollywood Vietnam War movies, the helicopter became the defining weapon of the conflict. Its visual impact and resulting iconic status became something of a cliché – nonetheless, this status was well deserved.

As a result of US involvement in Southeast Asia, helicopter deployment and tactics reached their height. The 1st Cavalry Division (Airmobile) alone could put 400 helicopters in the air over Vietnam. The Bell UH-1 Iroquois, better known as the Huey, was the robust, utilitarian workhorse of US forces. Its brave pilots constantly landed and picked up troops from 'hot' landing zones.

The Huey first arrived in early 1962 and within two years there were 250 of them serving in Vietnam. Helicopter pilot Warrant Officer Robert Mason recalled, 'My first impression of the machine was that it was pure silk … At operating speed there was no roaring, vibrating or shaking, just a smooth whine from the turbine.' Compared to the H-19, which he was trained on, it was a Rolls-Royce. Also, the Huey was to prove to be incredibly rugged and could take a lot of punishment, both in terms of handling and from enemy fire.

The short-fuselage models comprised the UH-1A, B, C, E and P, while the longer bodies were designated the UH-1D, F, H and N. The UH-1D was first delivered in 1963, but was superseded in combat units by the H model. The Huey was deployed as a gunship (guns), transport (slick) and air ambulance (medevac). Controversially, the Huey was also employed to spray jungle defoliant. Along with the door-mounted M60 machine gun, gunships were also armed with 40mm grenade launchers and 2.75in rockets.

Robert Mason observed, 'Our two door gunners would be stationed on either side of the hell-hole-cover – in the pockets – firing M60 machine guns attached to pylons. During our first two months, though, the machine guns would simply be strung from the top of the open doorways on elastic bungee cords. With the crew chief and gunner in the pockets, there was enough room for eight or ten troopers on the cargo deck.' Crucially, in the early days the pilot and co-pilot lacked armoured chest plates that could protect them from Viet Cong and North Vietnamese Army small arms fire.

The helicopter's first major conventional war was in fact Korea, where such aircraft as the Sikorsky S-51/H-5 and Bell 47 were used for casualty evacuation (casevac) and reconnaissance purposes. In Indochina the French first deployed American-built H-19B helicopters against the Viet Minh in 1954, although this was too late to help stave off defeat. France, determined not to lose Algeria as well, quickly became one of the world's leading authorities on the use of helicopters in combat conditions, after buying further aircraft from the US. These included the Boeing Vertol OH-21 Work Horse/Shawnee. In order to protect these slow transports from the Algerian National Liberation Front, the French Army's 2nd Helicopter Group experimented with mounting machine guns on its aircraft for counter-insurgency operations.

Whilst the helicopter was used in air assault operations in Korea, Suez and Algeria, the concept of massed helicopter air mobility was not fully appreciated until American intervention in Southeast Asia. The Vietnam War was where the helicopter truly came into its own, playing a significant part in the bloody and protracted conflict. Although the helicopter proved its utility in a whole host of roles, the war highlighted its extreme vulnerability to small arms fire and ground fire in general. This was attested by the very high helicopter casualty rates, which have never been matched since.

In December 1961 former escort carrier, USNS *Card*, arrived off Saigon with thirty-two Vertol H-21 Shawnee helicopters. The following year the first Hueys were initially deployed for supply purposes, but quickly became pressed into service to ferry troops into combat. By 1964 more than 400 US Army aircraft had been deployed in theatre, including 250 UH-1 Hueys and nine CH-37 Mojave heavy transport helicopters. The CH-46 Chinook medium-lift helicopter also earned its stripes in Vietnam.

The first US Army airmobile unit to be committed was the 173rd Airborne Brigade in June 1965, followed by the 1st Cavalry Division (Airmobile). The latter incorporated elements of its predecessor, the 11th Air Assault Division (Test) from Fort Benning, along with units from three other divisions. The 1st Cavalry, stationed at An Khe on Route 19, regularly had hundreds of helicopters airborne at any one time. Pilots were taught to fly in very close formation and at low level so that a standard flight of four Hueys could get into a small landing zone.

As the war progressed, the landing zones were increasingly contested. 'I could see those tracers, in their lazy-looking flight upward, from five miles away,' recalled WO Mason. 'In between each tracer were four more bullets. A fifty-calibre machine gun spits out bullets a half inch in diameter and an inch long … When blasted out a gun at 3,000ft per second, it had incredible power and range.' These rounds easily punctured the fuselage and the pilot's Plexiglas.

In 1968, crucially North Vietnamese General Giap underestimated the technological strides that had been made since Dien Bien Phu when he launched his Tet Offensive. The helicopter and transport aircraft sustained a lifeline to the US firebase at Khe Sanh, and indeed air supply delivered 12,000 tons of cargo. In addition, more than 600 paradrops were made, which were vital for those men on the exposed hills. On the ground, the communists did ensure that no convoys got through by keeping up their artillery attacks on American bases at Dong Ha, Con Thien and Camp Carroll.

Although Tet was largely over by 26 February 1968, Khe Sanh remained surrounded. A month after the Tet Offensive was thwarted the US 1st Cavalry was sent to lift the seventy-seven-day siege with Operation Pegasus/Lam Son 207. The Americans and the Army of the Republic of Vietnam (ARVN) put together a relief force, numbering 30,000 troops. The 1st Cavalry landed 10 miles from Khe Sanh on 1 April 1968, and there they joined other American and ARVN forces to open Route 9. Elements of the Air Mobile entered Khe Sanh largely unopposed six days later.

Military clear-up activity in the area continued well into the following year. Operation Scotland II ran until February 1969 and resulted in more than 3,000 enemy casualties around Khe Sanh. The base itself was evacuated in mid-1968 but was temporarily recaptured by the ARVN in 1971. After that it remained in North Vietnamese Army (NVA) hands and by 1973 the airstrip had been turned into an all-weather MiG fighter runway.

By 1968 the South Vietnamese Air Force possessed about seventy-five H-34 helicopters, and by the end of 1972 it had some 500 new machines including Hueys; one of the largest, costliest and most modern helicopter fleets in the world. Just three years later, with the South's collapse, all were destroyed or had fallen into North Vietnamese hands.

The largest, longest-ranging combat helicopter assault was Operation Lam Son 719 on 6 March 1971, when 120 Hueys ferried two ARVN infantry battalions 77km into

Laos. They only lost one helicopter on the approach. This impressive success was marred by the complete shambles of the evacuation and by the fact that, during Laotian operations between 8 February and 9 April 1971, the Americans lost 108 helicopters with another 600 damaged. All this helped hasten America's departure.

The first instance of a tank being destroyed by a guided missile-firing helicopter occurred on 2 May 1972, during North Vietnam's Easter Offensive. Two UH-1B knocked out a NVA light tank using TOW anti-tank missiles, and in subsequent actions went on to destroy another twenty-six tanks. There was little real novelty in this as helicopters and guided anti-tank missiles had been in existence since the Second World War. However, it had taken almost three decades to successfully mate the two technologies, and now the anti-tank helicopter is considered to be one of the most dangerous opponents of armoured fighting vehicles and fixed positions.

Ultimately the US forces' and ARVN's dependence on the helicopter was to prove a fatal flaw – especially after the NVA became equipped with the Soviet man-portable SA-7 anti-aircraft missile (effective against all subsonic aircraft below about 3,000m). It also ran contrary to the hearts and minds campaign by distancing the military from the indigenous population it was supposed to be protecting.

Whilst Vietnam established the helicopter's indispensability as a battlefield transport and aggressor, it also highlighted its vulnerability. Staggeringly, more than 16,000 of all types were lost to enemy action or accident in Vietnam (though this equated to only one per every 4,000 sorties). Nonetheless, many were recovered and repaired, so that total combat losses amounted to 2,076, whilst accidents accounted for another 2,566.

In Vietnam, the UH-1 received the brunt of things. In total about 16,000 Hueys were built, with about 7,000 of them seeing action in Vietnam; of these 3,305 were lost along with 2,177 pilots and crew. Remarkably, this highly successful helicopter was used by more air forces and built in greater numbers than any other military aircraft since the Second World War.

Vietnam was an incredibly high-tech war, particularly in the air. Helicopters and transport aircraft were employed to spray chemical defoliant to destroy huge swathes of jungle canopy. Both were also used to drop the massive 10,000lb Daisy Cutter bomb. This exploded 3ft above the ground, and the blast, which radiated outwards, flattened jungle, and destroyed enemy bunkers and minefields all in one go. These cleared areas created instant helicopter landing zones. A similar weapon was the Fuel Air Explosive, which could clear jungle to a diameter of up to 30m. At the lower-tech end of things, Hueys were employed as bombers to drop mortar rounds on enemy targets. They were also fitted with loudspeakers to broadcast propaganda.

Helicopters and transport aircraft were converted into lightships that could turn night into day in support of ground operations. During darkness transport aircraft converted to heavily armed gunships could pounce on unsuspecting targets. In the sky, surveillance aircraft patrolled constantly, monitoring enemy radio chatter and intercepting sensor intelligence. The Americans also used surveillance drones, based on jet-powered aerial targets, and unmanned aircraft in Vietnam. Notable amongst the latter was the Pave Eagle, a remote-controlled version of the civilian Beech Debonair aircraft. This was used to relay signals from air-delivered seismic detection sensors dropped along the Ho Chi Minh trail.

13

NAM 'MUD-MOVERS'

American fighter-bombers also played an important role in the conflict in Vietnam. The North American F-100 Super Sabre was employed primarily as a 'mud-mover', delivering unguided rockets and bombs. It became the first jet to drop ordnance in Southeast Asia on 9 June 1964. The Super Sabre's role rapidly expanded and by 1967 it was the most numerous fighter in South Vietnam. Both single-seat F-100Ds and two-seat F-100Fs also served in Laos and to a lesser extent Cambodia and over the North Vietnamese panhandle. The Super Sabre was affectionately dubbed the 'Hun' by simply shortening its numerical designation.

In the early stages of American involvement in combat missions the 'Hun' initially flew sorties over North Vietnam and Laos. However, its performance was not sufficient for operations over the heavily defended North, so it was primarily used for in-country missions over the skies of South Vietnam. Most of these were typically TIC sorties, where the 'Huns' supported 'troops in contact' with the Viet Cong and North Vietnamese Army. The 'Hun' conducting close air support was always a very welcome sight for US and South Vietnamese forces pinned down by enemy fire.

The F-100, which first came into service in the mid-1950s, was designed as a single-seat tactical fighter powered by a Pratt and Whitney J57-P-21A turbojet engine. It was capable of up to 864mph at 35,000ft and had a climb rate of 16,000ft a minute. The combat range was 1,500 miles. Between 1955 and 1959, a total of 2,294 Super Sabres were built as a successor to the North American F-86 Sabre, which had seen extensive combat in Korea. The Super Sabre was armed with four 20mm M-39 cannon (reduced to two on the F-100F) and could carry up to 7,500lb of external stores, including air-to-air and air-to-ground missiles, rockets and bombs. This made it an ideal fighter-bomber.

In the early 1960s, as the US Pacific Air Forces (PACAF) made ready to help the South Vietnamese, the organisation was singularly ill-prepared for the task. It was configured to fight a major conventional war with a reliance on nuclear deterrence. PACAF's twenty-eight squadrons deployed some 600 aircraft, including 200 F-100D/F tactical bombers, from twelve bases in Hawaii, Japan, Korea, Okinawa and the Philippines. However, these units were not trained in counter-insurgency warfare.

It meant that Osan air base in South Korea was the only PACAF base on the Asian mainland from which jet fighters were operating regularly. As a result, the US began to build bases in Thailand and South Vietnam. When the communists began to step up their attacks in Laos and Vietnam, the USAF's presence was boosted with rotational deployments to Thailand. In December 1960 six F-100Ds of the

510th Tactical Fighter Squadron were sent to Don Muang air base as part of Project Bell Tone I.

It was not long before F-100s were deployed to Tan Son Nhut airport outside Saigon, the South Vietnamese capital. On 9 June 1964, eight F-100Ds specially deployed from Clark air base, in the Philippines, to Da Nang, South Vietnam, and, supported by RF-101C pathfinders and KB-50J tankers, flew a retaliatory strike against communist aircraft bases. This was to avenge the loss of two US Navy planes shot down during a reconnaissance flight over Laos. Two months later, the Gulf of Tonkin Incident sparked massive USAF involvement in Southeast Asia.

While retaliatory raids were limited to carrier-based air-craft, the USAF began to boost its presence in the region. The F-100Ds of the 615th Tactical Fighter Squadron at Da Nang were reinforced by ten F-100Ds of the 405th Tactical Fighter Wing sent to Takhli in Thailand. They were involved in operations in early February 1965 when they struck communist anti-aircraft sites. The Super Sabres first arrived in theatre without tactical camouflage. A growing anti-aircraft threat meant that it was not long before the aircraft were in the paint shop.

Phase I of Operation Rolling Thunder (attacks against enemy lines of communication in the panhandle of North Vietnam below the 20th parallel) commenced on 2 March 1965. This included forty F-100D/Fs, although the fleet of Super Sabres was soon to expand to seventy-two aircraft. By the end of 1965 there were four squadrons of F-100s – the 308th, 510th and 531st TFS at Bien Hoa and the 416th TFS at Tan Son Nhut.

Flying four F-100Fs specially fitted with the Radar Homing and Warning System (RHAW), Detachment

1 of the Tactical Warfare Center arrived at Thailand's Korat air force base on 26 November 1965 to evaluate vital surface-to-air missile countermeasures. Employing RHAW, the Wild Weasel F-100Fs were designed to warn strike aircraft of impending surface-to-air missile (SAM) launches and to home in on the SAM's Fan Song guidance radar to direct F-105s assigned to the Iron Hand SAM suppression missions.

Eventually, in order to attack the SAM sites themselves, the F-100Fs were armed with AGM-45 Shrike missiles. Starting in May 1966, Wild Weasel operations were conducted by modified EF-105Fs and from September 1970 the more capable F-105Gs. The F-100F 'Misty FAC' aircraft was a two-seat 'Hun' that flew forward air controller missions at low level and high speed. As the air war escalated, the North Vietnamese Air Force rose to meet the Americans and losses on both sides began to mount. During 1966, the USAF lost 379 aircraft, a third of which were F-105s that were mainly lost to ground fire; this total, though, also included twenty-six F-100s.

Evaluation of the new Northrop F-5A Tiger fighter resulted in its being transferred to the South Vietnamese Air Force. The F-102A also proved to be of limited use because it was not suitable for ground support. The B-57 was not available in sufficient numbers, so the F-100 and the F-4 became the most numerous fast jets in South Vietnam. In particular, the F-100 was ideal as its performance was suited to in-country operations where there was little enemy air opposition. At the beginning of 1967 some eleven tactical fighter squadrons were flying 198 'Huns' from bases in South Vietnam; seven squadrons flew 126 F-4Cs and one squadron was still equipped with F-5As.

The F-100D Super Sabres of the 37th TFW became operational at Phu Cat air base on 29 May 1967. The initial squadron was the 416th TFS, which was then joined by the 355th and then the 612th. The 'Huns' presence in South Vietnam peaked in 1968 when the Air National Guard (ANG) provided four additional squadrons of F-100Cs plus volunteers to support a USAF F-100 unit. When the ANG units were activated, the twenty-two F-100s of the 174th TFS, Iowa ANG, joined the 37th TFW at Phu Cat. The huge 37th TFW had up to 110 F-100s, including eighteen F-100F 'Misty FAC'.

As well as the 3rd, 35th and 35th TFWs, the massive F-100 force in Southeast Asia (some 490 aircraft) included the 31st TFW based at Tuy Boa from mid-1967. This wing included the 308th and 309th TFS as well as the New Mexico ANG's 188th TFS (known as the 'Enchilada Air Force') and the New York ANG's 136th TFS ('Rocky's Raiders'). Tuy Boa on the coast was dubbed 'The Atlantic City of the South China Sea'. Fuel supplies for the base had to be piped 15 miles overland from Vung Ro Bay. Every now and then the Viet Cong would blow the pipeline and the F-100s would bomb them in retaliation.

In February 1967 the Americans opened Operation Junction City, the only parachute assault in the whole war. This was supported by F-100s and naval air strikes. On 21 March 2,500 communist troops attacked the US fire base in the Michelin rubber plantation 20 miles north-east of Tay Ninh. The 'Huns' were called in from Bien Hoa and Phantoms from Camh Ranh Bay. The F-100s and F-4s flew more than 5,000 sorties supporting Junction City and the Viet Cong and NVA were eventually driven back into their Cambodian sanctuaries.

The 'Huns' were kept very busy throughout the Vietnam War, to the extent that on occasions at Bien Hoa an F-100 landed or took off every forty-two seconds. In 1969 alone the F-100s conducted 52,699 sorties. After flying more than 360,000 combat missions, the last Super Sabres operating in Southeast Asia belonging to the 35th Tactical Fighter Wing ceased operations at Phan Rang air base in June 1971. During the war the USAF lost 186 F-100s to anti-aircraft artillery, seven on the ground as a result of air base attacks and forty-five in operational accidents. The Super Sabre was finally retired from service with the US ANG in 1979, some four years after the conflict in Vietnam came to an end.

14

DOWNTOWN HANOI

Vietnam enabled the US Air Force to practise strategic bombing techniques on a grand scale and develop vital electronic countermeasures to protect its bombers. By the early 1960s American involvement in Vietnam was gathering apace. The USAF implemented Operation Flaming Dart on 7 February 1965, which sent F-105 Thunderchiefs to attack targets located in North Vietnam. These air strikes were in reprisal for Viet Cong attacks on US bases in South Vietnam. The Viet Cong reacted with yet more attacks, prompting Flaming Dart II involving air strikes from three US carriers. The following month, 3,500 US Marines were sent to protect USAF bases in South Vietnam, and Operation Rolling Thunder began a sustained aerial bombardment of North Vietnam that lasted more than three years.

The American B-52 Stratofortress strategic bomber operated from bases on Guam Island and in Thailand during the Vietnam War. The Operation Arc Light raids first hit Viet Cong jungle hideouts in mid-1965 onwards. The B-52F was the first model employed, but it was the B-52D with the 'Big Belly' modification enabling it to carry up to 108 bombs that became the main heavy bomber of the war. The B-52 also conducted sorties over South Vietnam in support of ground troops. A total of 741 B-52 sorties were flown over the North and another 212 over the South.

The B-52 became one of the weapons that the Viet Cong and North Vietnamese Army most feared. In the wake of the communists' 1968 Tet Offensive and the heavy fighting at Hue, bombing sorties were increased from 800 to 1,200 monthly. On 15 February 1968 this was increased to 1,800. This meant that North Vietnam had to make air defence a priority.

The North Vietnamese first requested the SA-2 SAM from Moscow in the mid-1960s in a bid to counter US high-altitude bombing. Understandably, the Soviets were reluctant to supply the SA-2 for fear it might fall into American hands – if this happened then the USAF would be able to develop countermeasures to jam the missile's radar. Moscow was keen not to assist Washington's efforts to make the U-2 spy plane invulnerable to the SA-2, or for that matter the B-52. At the same time, the Soviets were keen to put the SA-2 through its paces and make the most of operating it against American aircraft – especially the fighter-bombers and bombers. Ultimately this would be a two-way intelligence bonanza.

The Soviet-designed S-75 Dvina high-altitude surface-to-air missile, code-named SA-2 'Guideline' by NATO, first went into production in the mid-1950s. It served with the Soviet Army at both front and army level, with eighteen launchers per air defence regiment. It became probably the most widely deployed SAM system of its kind in the world. During the Cold War the SA-2 was used by all the Warsaw Pact armies and was also widely exported to such countries as China, Egypt, India, Iraq, North Korea and Vietnam. In the dark days of the Vietnam War, the pilots and aircrew of the United States Air Force and the South Vietnamese Air Force learned to develop a healthy respect for the Soviet SA-2 SAM.

The missile first came to public notice after shooting down a number of high-altitude American U-2s operating over the communist bloc. This included Francis Gary Powers' aircraft over the Soviet Union in 1960. A total of fourteen missiles were fired at Powers, one of which hit his aircraft while another accidently struck a pursuing MiG-19 fighter. SA-2s supplied to China were also used to shoot down five Taiwanese-piloted U-2s. Then, in 1962, Rudolf Anderson and his U-2 were shot down over Cuba.

Subjected to continuous improvements, the 10.7m-long SA-2 missile has four sets of stabilising fins. These include cruciform small fixed nose fins, a second set of larger cruciform cropped delta wings two-thirds of the way along the body and a third smaller set toward the rear. In tandem was a solid boost motor with four very large delta fins, which like the others are indexed in line. One opposite pair has trailing edge controls for the initial roll stabilisation and gathering onto the guidance beam. The missile has a range of some 50km.

While the SA-2 was designed as forward deployable and fully mobile, it could not be fired on the move. It had to be deployed to fixed or temporary sites using a rotating launcher platform, which was elevated to about 80 degrees before firing. A single missile was transported on a ZIL-157-hauled articulated trailer, and from this it was pulled backwards onto the launcher. The standard support radar for the SA-2, known to NATO as the Fan Song, locked onto the target and fed data to the computer van. The latter set up the launcher to fire the missile and after firing employed an ultra high frequency (UHF) link to guide the missile once it had the guidance beam. It was this requirement for fixed launch sites that made the SA-2 very easy to detect during the Cold War.

It was the fixed sites that soon tipped off Washington that Moscow had agreed to Hanoi's request for the missile. In the spring of 1965, US intelligence officials received aerial photos of North Vietnamese surface to-air-missile and anti-aircraft artillery positions. In support of these intelligence gathering operations, America deployed assets such as the carrier-based RA-5C Vigilante, RF-101 Voodoo and RF-4C Phantom, the latter proving to be the most effective tactical reconnaissance aircraft of the war.

America detected SA-2 launch site preparation almost immediately when US reconnaissance aircraft photographed construction work on 5 April 1965. Almost four months later the SA-2 achieved its first kill over Vietnam when it shot down a USAF RF-4C on 24 July 1965. Washington did not take this lying down and immediately launched Operation Iron Hand to destroy the SA-2 sites before they became fully operational. However, for political reasons Hanoi, the North Vietnamese capital, was considered out of bounds at that stage. Such operations to destroy SAM sites and their supporting radars were later dubbed Wild Weasel missions.

From 1967 to 1972 there were some 200 SA-2 missile sites in North Vietnam, but by 1972 around 150 missiles were fired for every aircraft brought down, so they were not terribly effective. The SA-2 was mainly restricted to defending North Vietnam's infrastructure and was not used as a field weapon. For example, during the siege of the US firebase at Khe Sanh the main threat to American aircraft came from anti-aircraft artillery deployed in the surrounding hills, not missiles.

The USAF's experience with the SA-2 meant the Americans were able to develop electronic countermeasures that could jam its radar guidance frequencies. This was to have a significant impact in later conflicts that involved the SA-2. The American Douglas EB-66E Destroyer aircraft was an Electronic Counter Measures platform that was vital in helping to keep losses to North Vietnamese SAMs to a minimum. Flying from bases in Thailand, this force was kept busy supporting raids until the advent of effective ECM carried by the striker force itself.

The Iron Hand/Wild Weasel suppression of enemy air defences missions needed dedicated weaponry that could target the SA-2 and its supporting radar systems. To this end the US Navy deployed the AGM-45 Shrike anti-radiation missile, which was developed to target hostile enemy anti-aircraft radar using the body of the AIM-7 Sparrow air-to-air missile.

Shrike attacks commenced in October 1965 supported by USAF EB-66 bombers equipped with radar jammers. Following this, Soviet technicians were able to upgrade the SA-2's resistance to electronic countermeasures. They also ensured that the Fan Song tracking radar could lock onto the jamming signal and send the missiles towards the jammers. Passive guidance on the SA-2 missile also meant that the tracking radar could be switched off to confuse the Shrike.

The North Vietnamese crews learned to point the tracking radar in the wrong direction. The Shrike would home in on the beam and then crash in the completely wrong spot as soon as the radar was switched off. They also developed a tactic of dummy launches whereby the Fan Song guidance signal was broadcast but the SA-2 was not actually fired. This bluff could cause an attacking aircraft to deliver its ordnance too soon. Throughout the war America's technical boffins were put to work to come up with ways of countering North Vietnam's countermeasures.

The SA-2 really came into its own defending Hanoi and Haiphong against Operation Linebacker II. On 18 December 1972 three B-52s were lost to SA-2s and an A-7 Corsair was also hit. Two days later they brought down or damaged six B-52s and an A-6 Intruder. On 21 December two more B-52s and another A-6 were hit by the SA-2. Over the next few days another four B-52s were brought down by the missile. In total ten B-52s were lost over the North while five others limped away only to crash in Laos or Thailand. Linebacker II cost the Americans a total of twenty-six aircraft, including fifteen B-52s. The North Vietnamese claimed they shot down thirty-four B-52s and four F-111s during the campaign.

Such losses seemed to indicate a clear victory for the SA-2, however this was not all the story. The Americans had kept developing their ECM techniques and when Hanoi was attacked during Linebacker II it took 266 missiles to hit just fifteen bombers. The North Vietnamese even resorted to firing salvoes of missiles but this was very wasteful. Throughout December 1972 the North fired a total of 1,242 surface-to-air missiles at the USAF, exhausting their defences and leaving the Americans free to roam at will. However, with Moscow's support, within two years the North had recovered its strength. This effectively neutralised South Vietnam's air superiority and by 1975 the North Vietnamese Army was superior in both weaponry and supplies, thus giving the North the technological edge.

For a long time the USAF largely had control of the skies over Southeast Asia. In total it claimed 137 North Vietnamese MiG kills. Nevertheless, by 1972, once the North Vietnamese Army became equipped with the shoulder-launched SA-7 SAM and radar-controlled anti-aircraft artillery, all but the most modern aircraft became obsolescent. It also restricted low flying and helicopter operations.

North Vietnam's anti-aircraft artillery defences were formidable and accounted for up to 68 per cent of US aircraft losses. However, North Vietnamese surface-to-air missiles only accounted for 5 per cent of US combat losses; this was because of the effectiveness of American electronic countermeasures and the Wild Weasel air defence suppression missions. Nonetheless, as far as North Vietnam was concerned, the SA-2 had done its job in the skies over Hanoi.

Canadians serving with the RAF scramble to their Hurricanes. In the summer of 1940 Hitler employed his Luftwaffe as a terror weapon to try and beat Britain into submission, heralding the Battle of Britain. (Crown copyright)

A shot-down Heinkel 111 medium bomber. This aircraft had been designed as a tactical, not a strategic, weapon. Nonetheless, it was used to attack London and other major cities. (*Portsmouth News*)

Hitler's attacks on the RAF and Britain's cities forced people to seek sanctuary in underground air raid shelters. (*Portsmouth News*)

The Luftwaffe also deployed its Ju 87 Stuka dive bombers during the battle but these soon proved vulnerable to the more agile Hurricane and Spitfire. (Via author)

During the Battle of Britain, German pilots and aircrew suffered heavy losses pressing home their air raids. (Author's collection)

On 7 December 1941 Japanese carrier aircraft surprised the US fleet at Pearl Harbor naval base, sparking war in the Pacific. (US Navy)

Japanese A6M2 'Zero' fighter on board the carrier Akagi. (US Navy/Robert L. Lawson Collection)

The US destroyers *Downes* and *Cassin* lay wrecked in front of the battleship *Pennsylvania*. Luckily for the US Navy, its carriers were not at Pearl Harbor. (US Navy)

Aircrew grapple with a Dauntless dive bomber on the flight deck of the carrier USS *Hornet*. (US Navy)

Douglas TBD-1 Devastator torpedo bombers on the USS *Enterprise* during the Battle of Midway. (US Navy)

Midway left the Japanese carrier fleet crippled. The *Hiryu* ended up dead in the water once the fires spread out of control, despite her efforts to escape westward. (US Navy)

The Hampden was one of the RAF's most important medium bombers at the start of the Second World War. (Via author)

The flawed twin-engine Manchester bomber was the predecessor of the much more famous four-engine Lancaster. This one is being loaded with 2,000lb bombs in April 1941. (Via author)

The Halifax flew alongside the Lancaster as part of RAF Bomber Command's strategic air offensive against the Third Reich. (Via author)

Bombs being loaded into a Halifax bomb bay. The 'Bomber Barons' were convinced they could force Hitler to surrender by destroying his factories and fuel supplies. (Via author)

The Focke Wulf Fw 190 proved to be a superior aircraft to the more famous Messerschmitt Bf 109. (Via author)

Allied bombers conducted operation Pointblank in the run-up to D-Day in order to cripple the German lines of communication. (US National Archives and Record Administration)

Typhoon fighter-bombers on patrol. Note the twin 20mm cannons on the wings. These, combined with its speed, in the right hands made it a deadly adversary for the Luftwaffe. (Courtesy Alan Jones)

There is no denying that the Typhoon was an impressive-looking aircraft. However, despite its tank-busting reputation it was scrapped after the war as it was dangerous to fly, vulnerable to flak and its rockets were inaccurate. (Via author)

In mid-August 1944 formations of Boeing B-17 Flying Fortresses attacked targets in southern France and Italy as a prelude to Operation Dragoon, while B-25 Mitchell and B-26 Marauder bombers raided the port of Toulon. (USAF)

A USAAF side gunner engages enemy targets. Open to the elements and vulnerable to enemy fire, this was not a good place to be, especially while over the target. (USAF)

The air war took on a new dimension in 1944 when Hitler started launching V-1 flying bombs and V-2 rockets. This piloted Kamikaze version was developed but never deployed. (Via author)

Desperate to escape its tormentor, a MiG-15's final moments are caught by gun camera over Korea. (USAF)

F-86 Sabres photographed in June 1951. Initially this aircraft flew with impunity, but the arrival of foreign pilots operating with the Chinese Air Force resulted in losses and hampered attempts to cut communist supply lines. (US DoD)

The helicopter's first major conventional war was Korea, where such aircraft as the Sikorsky S-55/H-19 (seen here) and Bell 47 were used for casualty evacuation and reconnaissance purposes. (US Army)

The North American F-100D Super Sabre was America's heavy-duty bomb carrier over Vietnam and flew more sorties than any other aircraft type. (USAF)

The rugged Bell Huey UH-1 came to define the Vietnam War. Hueys acted as transport and medevac helos, while gunships provided vital support for the ground troops. (US Army)

Hueys dusting off after dropping their ground troops. (US Army)

The Soviet SA-2 surface-to-air missile came into its own resisting the US B-52 bomber over North Vietnam. (USAF)

North Vietnamese SA-2 crew celebrating a victory – in 1972 they hit fifteen B-52s. (NARA)

North Vietnamese anti-aircraft artillery actually claimed more American aircraft than surface-to-air missiles. (NARA)

The formidable Soviet Mi-24 helicopter gunship first made its mark in Afghanistan during the 1980s. (Via author)

An RAF Puma lifts off from Bessbrook, while an Army Air Corps Lynx conducts its pre-flight checks. Rapid response was vital in the war against the IRA in Northern Ireland. (UK MoD)

The RAF's Tornado has seen action in numerous wars, including Operation Desert Storm in 1991. (British Aerospace)

The US F-117 Night Hawk was the first aircraft to use low-observable stealth technology. It first saw combat in Panama in 1989 and was then deployed in Iraq two years later. (Lockheed)

A British Army Longbow Apache unleashing its deadly rockets over Afghanistan against Taliban targets. (UK MoD)

In 2011 NATO airpower helped topple Colonel Gaddafi in Libya by supporting a popular rising against him. This included Canadian F-18 fighters. (Canadian Air Force)

The development of the armed MQ-1 Predator unmanned aerial vehicle has led to a new facet of aerial warfare – the drone wars. (USAF)

Pilots no longer have to fly with their aircraft as armed drones are remotely operated. (USAF)

The MQ-9 Reaper, a larger version of the Predator, has been used to kill terrorists in Pakistan and Yemen. (USAF)

15

NIAGARA

The North Vietnamese communists launched a massive co-ordinated surprise attack – the Tet Offensive – across South Vietnam against the South Vietnamese and American armies on 30 January 1968. Despite many of its personnel being on leave for the Tet holiday, the South Vietnamese Air Force (VNAF) reacted swiftly, with those aircrews on duty bearing the brunt of the opening air operations until their comrades returned.

The VNAF got as many aircraft into the air as possible, throwing a lifeline to the hard-pressed Army of the Republic of Vietnam (ARVN). During late January this included 215 sorties by fighter aircraft, 196 by reconnaissance aircraft, 158 by transport aircraft and 215 by helicopters. During February these sortie rates increased, with the VNAF strafing and bombing the North Vietnamese Army and Viet Cong guerrillas (NVA/VC) throughout South Vietnam.

The Cessna A-37B Dragonfly and the F-5E Tiger had been used to bring the VNAF up to strength and create an all-jet fleet. In August 1967 about ten A-37s arrived at Bien Hoa and Pleiku for evaluation by the United States Air Force's (USAF) 604th Fighter Squadron under project Combat Dragon. The American A-37s were soon joined by Vietnamese-flown A-37s of the VNAF 524th Fighter Squadron at Nha Trang. The highly versatile A-37 was used for all light strike missions, including counter-insurgency, forward air control and rescue escort.

Vietnam saw one of the most intensive air wars fought over a ten-year period. No other air war, with the exception of Korea, has come anywhere near the massive number of sorties conducted by the USAF over Vietnam. In particular, the bitter engagement fought for the US Marine fire base at Khe Sanh, in South Vietnam, as part of the Tet assault saw tactical airpower coming of age.

The American base at Khe Sanh experienced one of the deadliest deluges of munitions unleashed on a tactical target in the history of warfare. Both the siege of Khe Sanh and the bitter battles of Hue and Saigon stretched American resources to breaking point. Although in each case victory was secured, the Tet Offensive was to have far-reaching ramifications for America's commitment to South Vietnam and the conduct of the war.

The USAF, US Navy (USN) and VNAF conducted an escalating bombing campaign against North Vietnamese targets under the code name Operation Rolling Thunder from 2 March 1965 to 2 November 1968. While this was the largest air offensive conducted since the Second World War, it was greatly hampered by inter-service rivalry and political restraints. In addition, the USAF was woefully ill-prepared for this type of operation, having spent decades planning for nuclear war against the Soviet Union. In contrast, the USN had the new A-6 Intruder all-weather fighter-bomber and was responsible for the F-4 Phantom, which became ubiquitous during the Vietnam War.

It was over Vietnam that the F-4 Phantom assumed the mantle of the P-51 Mustang of the Second World War and the F-86 Sabre of Korea. It accounted for 107 enemy fighter aircraft (and one kill shared with an F-105). During 1965–67 four MiGs were shot down for every American fighter lost. However, the Americans were nowhere near the ten to one ratio established during the Korean War.

Thuds, or Republic F-105 Thunderchiefs, bore the brunt of the air war against North Vietnam when Rolling Thunder first commenced. They were based in Thailand, where there was supposed to be an agreement preventing them operating over South Vietnam against the Viet Cong.

In part Rolling Thunder was intended to deter Hanoi from supporting the war in the south by cutting its supply routes, along which flowed equipment and men. Instead, North Vietnam greatly strengthened its air defences, taking a deadly toll on its attackers, and continued to supply the NVA/VC. There was also a hope that the air attacks would force Hanoi to the negotiating table.

The Tet Offensive was a nasty wake-up call for the USAF, USN and VNAF that showed their efforts had been largely for nothing. In early 1968 all available aircraft were diverted to support Operation Niagara during the siege of Khe Sanh. Likewise, B-52 tactical bombing raids were also deployed in support of the US Marines at Khe Sanh.

In the spring of 1965 the USAF was opposed, first by MiG-17s and later by MiG-19s and 21s. Some of the USAF's first air-to-air combat losses occurred on 4 April 1965 when four North Vietnamese MiG-17s jumped a force of F-105 Thunderchief fighter-bombers and F-100 Super Sabres, scoring two kills. Then, on 17 June 1965, two F-4 Phantoms intercepted four MiG-17s, claiming two with radar-guided Sparrow missiles.

Three days later, four piston-engine A-1 Skyraiders took on two MiG-17s and shot one down with 20mm cannon fire. From 1965 onward the North Vietnamese Air Force (NVAF) mainly operated the MiG-17F clear weather interceptors, as well as some Chinese-built Shenyang F5s (the export version of the J-5). It also received some MiG-17PFs and MiG-17PFUs. During 1967–68 the NVAF suffered heavy losses and for all practical purposes ceased to exist.

The Tet Offensive caught General William C. Westmoreland and the American military off guard with an intelligence failure ranked alongside Pearl Harbor. The NVA and VC assault effectively started in September 1967 when communist forces launched attacks against the isolated American garrisons in the central highlands. With great foresight, Westmoreland warned Washington in late December 1967 that he expected the communists 'to undertake an intensified countrywide effort, perhaps a maximum effort, over a relatively short period of time'.

The Vietnamese communists had stepped up their activity by early January 1968 with an attack on Da Nang air base, destroying twenty-seven aircraft. Attacks were also made in the Que Son valley and against the Ban Me Thuot airfield and An Khe. The VC-announced seven-day truce for the Tet holiday was schedule to begin on 28 January and VC activity died down the day before, apart from shelling of Khe Sanh.

American air power unleashed during the Tet Offensive was absolutely devastating. US naval air strikes were used in support of the Marine defenders of Khe Sanh in early 1968; 3,100 sorties were flown by US-carrier-based aircraft alone during February and March. US air support for Khe Sanh also included more than 7,000 sorties by the 1st Marine Aircraft Wing. In total, more than 24,000 tactical sorties and 2,500 B-52 bomber sorties were launched against the surrounding communist forces. In total, a staggering 100,000 tonnes of bombs were expended.

In particular, the B-52s dropping 75,000 tonnes of munitions over nine weeks provided the heaviest firepower ever unloaded on a tactical target. General Giap himself almost became a victim of the B-52s when he visited Khe Sanh in late January 1968. Operation Niagara even initially included the possibility of dropping tactical nuclear weapons.

In response to the Tet Offensive, the VNAF flew 7,213 sorties over four weeks, during which time it expended 6,700 tonnes of ammunition as well as transporting 12,200 men and 230 tonnes of supplies and equipment. The VNAF lost seventeen aircraft; ten on the ground and seven in the air, the latter comprising five A-1s, one C-47 and one U-17 (a Cessna 180 derivative). Overall, despite its relatively small size, the VNAF fought well.

At the height of US involvement in Vietnam a daily average of 800 sorties were flown by US fighter-bombers in support of the ground forces. In a single year four American aircraft types alone flew well over 80,000 sorties.

The USAF dropped more than 6 million tons of ordnance in Southeast Asia, more than double that used in the Second World War and Korea combined.

Following Tet it was only the halt to American bombing missions over North Vietnam on 1 November 1968 that enabled North Vietnam to rebuild its air force. They were to receive some MiG-19s, which were mainly the Chinese-built copy, the Shenyang F6 (export version of the J-6). In mid 1966 North Vietnam had about sixty-five fighters (most were MiG-17s, although they also included a dozen MiG-21s); by early 1972 this force had expanded to 200 (half of which were MiG-21s).

America's use of what later became known as 'shock and awe' to defeat the Tet Offensive horrified the American public, many of whom had no stomach for defeating the communists at any price. Despite American ingenuity and technology, the use of overwhelming air power proved to be a blunt instrument that targeted not only enemy soldiers but also innocent civilians. The widespread use of napalm and Agent Orange stripped the US military of any moral high ground. Great swathes of Vietnam became a poisoned wasteland. Inevitably this sapped America's will to prop up South Vietnam.

16

BANDIT COUNTRY

Britain turned to the helicopter to help with security operations across the British Empire in the 1950s and 1960s; in particular it faced down the communist threat in Malaya. Much closer to home, the helicopter later played a valuable role in countering the Irish Republican Army (IRA) in Northern Ireland during The Troubles. This placed the helicopters of the RAF and the Army Air Corps (AAC) directly in the firing line.

Above the deafening whine of the rotors, the pilot and his co-pilot probably did not hear the 'clunk-clunk-clunk' of the machine gun firing in at them. What they must have felt was the impact of the heavy rounds striking the side of their helicopter. Known as the 'Dashika' by the Mujahideen, the Soviet 12.7mm DShKM heavy machine gun saw extensive action with the rebels in Afghanistan against Soviet helicopters flying nap-of-the land operations during the 1980s. The IRA tried to replicate the

Mujahideen's success by attacking a British Army helicopter with such weapons in Northern Ireland in 1990.

The Soviet- and Chinese-made versions of the 'Dashika', along with the 'Ziqroiat' ZPU-1/ZGU-1 14.5mm, was the Mujahideen's standard air defence weapon in Afghanistan. Fortunately for the British armed forces, the IRA proved fair less competent than the Mujahideen in downing helicopters. Surprisingly, helicopters in Northern Ireland remained largely immune to ground fire; while a number were hit over the years, only a few were forced to land due to the damage.

From 1970 onwards the British armed forces deployed to Northern Ireland under Operation Banner were faced by a plethora of extremist Catholic Republican and Protestant Loyalist organisations bent on bloodshed. Central amongst them was the IRA (split between the Official and the Provisional/PIRA) and the Ulster Defence Association. Initially, the Army's primary role was to contain and then curtail the escalating sectarian violence between the Republicans and the Loyalists. As law and order was re-established in the main urban and rural areas, the popularity of the PIRA began to decline.

For security purposes, Northern Ireland, or Ulster, was split into three to match the police regions: 3rd Brigade was put in charge of Armagh and Newry (south and east); 8th Brigade covered Londonderry and Enniskillen (north and west); and 39th Brigade had responsibility for Greater Belfast northward to Larne. Once the rival communities were kept apart, the Army was able to move over to countering the terrorist threat. In response, the PIRA dug in, reorganised on a cell basis and prepared for a long war intent on uniting the North with the Republic of Ireland.

When the Army regained control in the key cities of Belfast and Londonderry, the Republican terrorists began to increasingly operate in the rural south from across the border in the Irish Republic. Sealing this border between the latter and Armagh was an impossible task. Building a secure border fence was politically unacceptable, so the Army had to settle for keeping a close watch on things. Foot patrols, which became the name of the game, were at constant risk from snipers, booby traps and remotely detonated bombs. These patrols had to disrupt the flow of terrorists and weapons across the border in order to prevent attacks further north.

The violent Catholic region of South Armagh became known as Bandit Country and, because of the mounting sectarian death toll, the Murder Triangle. This area was a traditional IRA stronghold so it was vital to defeat it on its home ground. The British Army established itself at Bessbrook, Crossmaglen, Forkhill and Newtonhamilton. Predictably, the IRA did not take kindly to this military presence.

When a Royal Artillery detachment deployed to Forkhill on a rural tour in the early 1970s, it found itself under persistent if inaccurate rifle fire from the nearby hills. A patrol was despatched and three firing positions were located. Back in Forkhill, a Bren gun was set up next to the unit latrines and it was zeroed in on the IRA positions. The men were instructed that every time they used the latrines they were to fire a burst in one of the three directions. This quickly had the desired effect and the sniping stopped.

Geography played a key role in dictating the use of helicopters. While they had little relevance in the densely populated urban areas of Belfast and Londonderry, they provided a lifeline for British Army posts in Ulster's rural areas, most notably South Armagh. Pushing convoys overland in Armagh eventually became so dangerous that the British Army began using helicopters to transport troops and supply its bases on a regular basis – a practice that was maintained until the late 1990s. The rolling landscape of Armagh was ideal helicopter country.

Certainly a number of Armagh bases, especially Crossmaglen in the very south of the county, were dangerous and hard to get to by road. The easiest and quickest way in and out of such places was by helicopter. Crossmaglen, known as XMG, had to be supplied almost exclusively by air as the logistical and manpower effort needed to mount a protected road convoy was not cost-effective. Also, the helicopter saved many lives in Northern Ireland conducting casualty evacuation. All the main hospitals in Ulster had helipads, or the injured could be ferried over to the UK mainland.

As The Troubles escalated so did the use of British Army and RAF helicopters in direct support of the counter-insurgency effort. Traditionally, the AAC provided direct aviation support for the Army, although it was also closely supported in this role by the RAF. The presence of fixed-wing Army aviation units in Northern Ireland was partly formalised in 1957 with the arrival at RAF Aldergrove of 1913 Light Liaison Flight equipped with Auster AOP.6s. This became 13 Flight of 651 Squadron within the revived AAC.

The Auster's successor, the de Havilland Beaver AL.1, became the only operational fixed-wing aircraft still in service with the Army (the Chipmunk was used for pilot training). It served the AAC in a liaison and communications capacity, but was also used in a general utility role ranging from taxi to ambulance. The last Beaver unit was based in Northern Ireland until the aircraft was phased out in the mid-1980s. The AAC also operated the Alouette AH.2 helicopter, which had been procured as a stopgap until the Scout became available in sufficient numbers. Its final days were spent in service with the British Army in Cyprus.

From 1970, nearly every Army brigade had at least one Aviation Squadron, usually consisting of twelve aircraft. The main rotor aircraft were the Scout and Sioux general purpose helicopters. These were soon bolstered by the introduction of the Westland Lynx in 1977 as well as the unarmed Gazelle. The Sioux was capable of carrying up to three passengers for observation and liaison purposes. Although a very basic helicopter based on an American design, it proved a first-class workhorse and was licence-built in Italy by Agusta.

The Westland Scout, if the rear seats were removed, could carry four soldiers facing out with their feet on the skids. This capability meant that the type was used to mount the first 'Eagle' patrols; the helicopter, having dropped off its soldiers, would then loiter ready to extract them. In most circumstances it was better for the helicopter to remain in the air than present a vulnerable target stationary on the ground. The first of 160 Westland Scout AH.1s took to the air in March 1961. In Northern Ireland the Scout was also used for intelligence gathering with the Heli-Tele, which provided real-time TV images to a ground-monitoring station; this was valuable for surveillance operations and crowd control.

In support the RAF provided the twin-engined Westland Wessex helicopter that could move up to sixteen men, depending on the distance covered. This twin-engine configuration gave it significant durability and lift ability. These were used to transport a Quick Reaction Force and were held on standby mainly in South Armagh. The Wessex, eventually replaced by the Puma, proved to be another invaluable workhorse throughout the 1970s in Northern Ireland. The RAF's heavy-lift Chinook also played a role.

In addition to the regular British Army, the eleven battalions of the Ulster Defence Regiment (UDR) were also trained to operate in RAF and Army helicopters. The UDR's tactical areas of responsibility covered much of Northern Ireland. The UDR, being the principal frontline unit from 1970, was effectively on permanent active duty (having been raised to shoulder the burden of security duties and relieve the Royal Ulster Constabulary). Its composition and role within the British Army was unique, with only a third of its manpower being full-time.

The Troubles gave birth to some unique military architecture. Across Northern Ireland, British bases were regularly attacked with home-made mortars and sniped at. The response was to create fortified camps complete with watchtowers, fencing and metal meshing to deflect the rockets. Helicopter pilots knew they could be subject to pot-shots by the IRA or hostile locals as they came in to land or took off.

To protect personnel, vehicles and the helicopters, bases such as Crossmaglen, which had a helipad near the barrack blocks, were surrounded by very high blast walls. AAC Lynxes, RAF Pumas and Royal Navy Sea Kings operated from Bessbrook's half dozen or so helipads, which were similarly fenced in to offer some protection from inaccurate missiles fired from beyond the perimeter. These were often launched off the back of construction lorries. Bessbrook at one point became the busiest helicopter airport in Europe.

The border posts resupplied by the helicopters were even more exposed to attack and this led to Northern Ireland's distinctive watchtowers. The Army watchtower at Camlough Mountain in South Armagh, for example, had an enclosed walkway running from the main tower down to two others and a barrack block. However, the helipad next to the walkway remained exposed.

By the late 1970s, the British Army in Ulster had received a new generation of helicopters, with the Sioux replaced by the Gazelle in the comms, liaison and recon role. Likewise, the Scout was replaced by the Lynx as the utility helicopter and in the support role the RAF's Wessex was superseded by the Puma. The RAF became one of the largest users of the Anglo–French Puma assault helicopter, which received forty of the original Puma HC.1s. The Gazelle was a French design and under a joint contract 135 Gazelle AH.1s were ordered for the British Army. The Westland Lynx performed several roles including tactical transport, armed escort, reconnaissance and evacuation. Unlike the Scout, the highly agile Lynx AH.1 was normally armed.

The requirements for rapid reaction forces led to the development of helicopter 'Eagle' and 'Duet' patrols.

The former were designed to insert foot patrols or surveillance teams. A Scout or Lynx would suddenly swoop in to allow troops to respond to incidents or establish instant roadblocks. Similarly, 'Eagle' patrols could quickly and discreetly retrieve men who had been on two-week surveillance operations. More complex operations could be conducted by the 'Duet' patrols that comprised an RAF Wessex or Puma and a Gazelle Command helicopter.

The Belleek Incident in mid-June 1976 during 3 Para's tour in South Armagh illustrates the utility of helicopters in countering terrorists in Northern Ireland. A local pub frequented by the PIRA was placed under observation and a hijacked car arrived at 9.40 p.m. Four well-armed men were subsequently pinned down during a firefight with a Para patrol. The terrorists barricaded themselves in a local bungalow, refusing to come out and taking the inhabitants hostage.

To stop them escaping, the company commander of the patrol company in a Gazelle and a Scout (with four men) was scrambled by Battalion Tactical HQ at Bessbrook; they were supported by another eight men under a platoon commander in a Puma. Arriving at 10.15 p.m., the company commander remained airborne to direct the action, ordering the Scout to land east of the inn to block the gunmen's line of retreat to the north-east; on arrival the Puma was instructed to come in to the south-west. Once the bungalow was surrounded and shots fired through a window, two of the gunmen quickly surrendered.

The Scout and Puma then flew back to Bessbrook to pick up a second platoon to help with the search for the others. The company commander supported the operation using the Gazelle and Scout's Nightsun searchlights to illuminate the ground and prevent the IRA men sneaking away under the cover of darkness. A third gunman was captured early next morning; he and the other two received fourteen-year prison sentences. The operation was lauded as a great success; the arrival of reinforcements and the use of the airborne command post had thwarted the terrorist plans and captured three of the four men and all their weapons, which included an assault rifle, a rifle and a sub-machine gun.

It was perhaps inevitable that such British helicopter operations could not continue with impunity. It was only a matter of time and sure enough on Friday, 17 February 1978 a British soldier was killed in a helicopter crash in Armagh. The IRA claimed to have shot it down but for many years the British Army denied this claim, before finally acknowledging that the IRA had indeed caused the crash.

Notably, the most effective IRA unit in attacking British helicopters during the conflict was the South Armagh Brigade. It reportedly conducted twenty-three separate attacks, reportedly forcing four down. The other successful IRA attack against a helicopter took place near Clogher, County Tyrone. It was the East Tyrone Brigade who managed to down this British Army helicopter using machine guns on 11 February 1990. It drew British troops to within yards of the Eire border and machine-gunned the unarmed helicopter, causing it to crash-land.

It is believed the brigade deployed at least two heavy-calibre weapons, possibly Libyan-supplied 12.7mm or American M60s, which fired 300 rounds at the AAC Gazelle. The two pilots were slightly hurt, and a sergeant major from the King's Own Scottish Borderers suffered spinal injuries in the crash. Only one previous aircraft, a Lynx, was acknowledged to have been brought down by gunfire, in 1988 in South Armagh. Allegedly it was the first casualty of its kind in the twenty years of conflict in the province (between 1969 and 1990 casualties included four RAF personnel).

The IRA also used a surface-to-air missile in 1991 against an AAC Lynx in Armagh, however it missed its target. To counter this new threat, British helicopters flew in pairs below 50ft or above 500ft. There is also evidence that the East Tyrone Brigade used missiles in attempts to shoot down British Army helicopters but to no avail.

The IRA imported large quantities of modern weapons and explosives from supporters in America and Libya. The IRA nearly obtained the US Redeye SAM from America in 1982. PIRA received 110 tons of weaponry from the Libyan regime of Colonel Muammar Gaddafi including machine guns, more than 1,000 rifles, several hundred handguns, rocket-propelled grenades, flamethrowers, SAMs and the plastic explosive Semtex in the mid 1980s. Reportedly, Gaddafi donated enough weapons to arm the equivalent of two infantry battalions. This brought the PIRA's new capability to the attention of the authorities on either side of the Irish border.

Libya supplied the Soviet-designed SA-7 SAM in revenge for UK bases being used in US raids on Libya in 1986. These missiles turned out to be out-of-date models and were unable to shoot down British helicopters equipped with anti-missile defences. Helicopters in Northern Ireland were equipped with infrared counter-measures such as flare dispensers (it was assessed that the

IRA would not target civil airliners as any American fatalities would harm US support).

While the prospect of the IRA or any of the other factions deploying SA-7 was publicly alarming, ultimately its success would have been a one-shot deal. When helicopters were restricted to 50ft or below, an SA-7 operator did not have time to acquire his target, fire and achieve a lock-on with the infrared heat-seeker. Ultimately the SA-7 would not have prevented helicopter operations unless it had been deployed in considerable numbers. Although the IRA possessed rocket-propelled grenades (RPGs), they never seem to have used them against helicopters, so there were no 'Blackhawk Down' type incidents. Tampering with the grenade's fuse may have been beyond the IRA's abilities.

By the mid-1980s, although the bulk of the AAC was based in Germany with the British Army of the Rhine (BAOR), a regiment-sized formation designated AAC Northern Ireland (ACC NI) was deployed in Ulster. This comprised three units: 655 Squadron with Gazelles and unarmed Lynx, the Beaver Flight with the Beaver AL.1 and a BAOR AAC Squadron on rotation. AAC NI was deployed at Aldergrove airport (north-west of Belfast) and Ballykelly (north-east of Londonderry). In 1993, the 5th Regiment AAC was created to support Operation Banner.

In 2002 it was announced that the RAF's Wessex helicopters would retire; in service since the 1950s, these included those stationed in South Antrim. That year a RAF Puma crashed in the county. Seven people were injured, two of them seriously, when the helicopter was forced to land on the side of a hill at Slieve Gullion. After almost forty years, Operation Banner ended at midnight on 31 July 2007, making it the longest continuous deployment in the British Army's history. Crossmaglen and most of the other bases were abandoned.

17

STINGING THE BEAR

By the 1980s surface-to-air missile technology was posing an increasing threat to fixed-wing aircraft and helicopters. In the case of Afghanistan, it helped hasten the end of Soviet intervention. The fighting there, like Vietnam before it, became dominated by air mobility, with Soviet troops developing slang for the different types of aircraft deployed there. Artyom Borovik and his comrades who served with the Soviet Army rather aptly dubbed the Mi-24 gunships 'bumblebees' and the Mi-8 transport helicopters 'bees'.

After Moscow's invasion of Afghanistan in 1979, the Soviet Air Force rapidly began to play a far greater role in the developing war against the resistance known as Mujahideen or 'God's Warriors'. By the summer of 1980, the Limited Contingent of Soviet Forces in Afghanistan were reorganising to meet the growing requirements of a protracted counter-insurgency war. At the end of the year there were about 130 Soviet fighter aircraft, mainly MiG-21s, Su-17s and MiG-23s, flying from Bagram, Shindand and Heart air bases.

This deployment also resulted in the commitment of new units with a large increase in the number of helicopters, rising from fifty to 300 by the following year. Three whole helicopter regiments were sent to Bagram, Konduz and Kandahar. Deployment of the *Vozduyushno Voorezhenie Sil* (VVS, consisting of the Strategic Air Armies, Air Force of the Military Districts and Groups of Forces and Military Transport Aviation, or VTA) was on a rotational basis, allowing most of the maintenance to be carried out in the Soviet Union. The main parent units were the 27th Fighter Aviation Regiment at Kaka and the 217th Fighter Bomber Regiment at Kirzyl Arvat.

By the second half of 1981, the Soviets were employing co-ordinated all-arms tactics backed by close air support (CAS). This emphasised concentration of air assets, extensive preliminary bombardment, then landing heli-borne forces to block and engage the enemy from unexpected directions, followed by a fully mechanised push usually toward a pre-positioned force. This was essentially the 'hammer and anvil' tactic.

An example of this CAS joint operation was conducted in August 1981, when government forces attacked the Mujahideen in the Marmoul gorge near Mazar-e-Shariff. A squadron of the 335th Air Regiment flew from Shindand to Dehdadi Air Force base near Mazar-e-Shariff, joining two squadrons of the Afghan 393rd Air Regiment. These Afghan units were supported by Soviet-based MiG-23s and spent a week bombing and strafing the Marmoul area. This was followed up by a reinforced battalion heli-borne landing by the Afghan 20th Division, but after heavy fighting it had to be airlifted out.

By 1985 there were ten Soviet aircraft squadrons in Afghanistan, with the same number in the Soviet Union flying support missions. There were two or three squadrons of MiG-23s and MiG-27 'Floggers', one or two squadrons of MiG-21 'Fishbeds' and two squadrons each of Su-17 'Fitters' and Su-25 'Frogfoots'. By 1984, the MiG-21 had been replaced by the MiG-23 and MiG-27.

Prior to the Soviet intervention, the Afghan Air Force (AAF) had about 169 combat aircraft. The Soviets then supplied the AAF with six MiG-21 'Fishbed' fighters, twelve Mi-24 'Hind' helicopter gunships: as well as a number of Su-20 'Fitter' fighter-bombers and Mi-6 'Hook' medium/heavy lift helicopters. This was the first occasion the AAF received the Mi-24 and Su-20; it also required an increase in Soviet advisers. After the Soviet invasion, the AAF (consisting of about seven fighter, helicopter and transport regiments) was co-located with Soviet units to avoid defection. In 1979–88 there were at least six defections by Afghan pilots to Pakistan with MiG-19s, MiG-21s and Mi-24s.

The primary Soviet weapon in Afghanistan was undoubtedly the helicopter. In particular the Mi-24 gunship was instrumental and the weapon most feared by the guerrillas. In 1982 the Soviets deployed up to 600 helicopters to Afghanistan, of which 200 were Mi-24s. They were used in preference to fixed-wing aircraft for most of the CAS missions, adopting nap-of the-earth tactics, especially once the Mujahideen's air defences became more sophisticated. Most of the troop-carrying and re-supply missions

were conducted by the Mi-8 'Hip', which was also capable
of an attack role, supported by the larger Mi-6. By 1983
the Soviets had deployed 150 Mi-8s and forty Mi-6s. The
ground-striking arm of the Soviet helicopter force was
provided by five tough air assault brigades.

The Soviet gunships were always swift to deal out retribution for attacks on heli-borne operations. When a
troop landing was attacked, Artyom Borovik, who had just
landed safely, witnessed:

> All four bumblebees have turned around to work a
> nearby cliff from which the *dukhi* [Mujahideen] have
> put our Mi-8 out of action. Their gunfire sends splinters
> of rock flying, enveloping the top of the mountain in a
> cloud of dust.
>
> For ten minutes more my chest throbs as the gun-
> ships, their glass sparkling in the sun, shoot at this and
> three other nearby cliffs.

Lieutenant Colonel Yuri Vladikin piloted a 'bumblebee'
flying combat support missions. He experienced just how
fearless the Mujahideen could be when they ambushed a
very large landing operation:

> For thirty minutes everything was quiet, but then even
> the stones began to shoot – that's how many emplace-
> ments there were. The bees are underneath; we are
> above. Those in the bees need real courage: the land-
> ing grounds are difficult to negotiate, and there are
> thousands of troops to land. Enemy fire comes from
> every cliff. I'm already circling the landing ground
> for a second hour. Very little ammunition left. We use

our resources sparingly and shoot only at large-caliber machine guns.

The number of helicopters serving in Afghanistan had fallen to 350 by 1985, but the Soviets still fielded 275 in February 1988, and the AAF around eighty organised into two regiments. The growing reliance on helicopters did impose operational restrictions. Engines overheated and were less efficient at high altitude, and there were problems with ice during the winter. Environmentally, this restricted efficient operations to the spring or autumn.

It was the Su-25 ground attack aircraft that proved to be the most effective bomber, with a high survivability against SA-7s. In April 1986, during the Zhawar campaign, Soviet Su-25s used laser-guided bombs to hit Mujahideen caves. Retard, RBK-250 cluster, laser-guided and even 12,000lb bombs were introduced, as well as high-altitude bombing.

Tu-16 'Badger' bombers were concentrated in the Turkestan Military District for high-level bombing of the Panjshir valley in support of the Soviets' seventh offensive in April 1984. The newer Tu-26 'Backfire' was first deployed into Afghanistan in November 1988. VTA An-12 'Cub' and An-26 'Curl' transport aircraft were used for reconnaissance and as master bombers. The older Soviet Il-28 'Beagle' bomber was deployed by the AAF's 355th Air Regiment at Shindand.

Before the arrival of significant numbers of SAMs (such as the SA-7 Grail via Egypt, Blowpipe via Nigeria and Stinger via Pakistan) the Mujahideen's primary anti-aircraft weapon was the tripod-mounted 12.7mm Dshk supplemented by the 14.5mm ZPU-1, ZGU-1 and the

23mm ZU-23 guns. Some Hinds were reportedly lost to SA-7s as early as 1980. Helicopters were also vulnerable to ground fire; during a sweep of the Panjshir in August 1981 the guerrillas brought down five.

In 1980–83 several dozen helicopters were lost to all causes, but an increasing use of flares by 1983 showed a growing fear of infrared heat-seeking missiles. Mujahideen in Kandahar shot down a Bakhtar Airlines Antonov reportedly using an American Stinger on 3 September 1985. The following year, the Soviets were becoming increasingly alarmed by the Mujahideen's use of SAMs, with aircraft routinely employing flares on take-off and landing.

In March 1986 it was clear the US had decided to supply Stinger to the Mujahideen. An initial batch of 200 was delivered in October 1986; by the time the US ceased supplies at the beginning of 1989, the Mujahideen had received about 1,000 missiles. The first effective use of Stinger was reported in the eastern province of Nangarhar. In the Zhawar campaign in 1986 the Soviets lost twelve helicopters and one fighter. During the first two weeks of November, eleven helicopters and one MiG-23 were reported shot down. By the end of 1986, the Soviets had lost a total of 500 helicopters, and by 1988 it was claimed Stinger had shot down 100 Soviet/Afghan aircraft of all types.

Artyom Borovik noted a rapid change in Soviet helicopter tactics:

Last year, the Mi-8s were flying at their maximum altitude – approximately six thousand meters. But now, with the appearance of the Stinger missiles, they descend from above and race along, just five meters or so off the ground, at a speed of two hundred and fifty kilometers per hour. They hover in the folds of the mountains, dart between the hills, and circle the kishlaks [villages] ...

Stinger also closed the airfield at Khost, which sparked off the Soviets' last major offensive of the war in December 1987 – in which heli-borne forces played a prominent role. Despite the US cutting off supplies in February 1989, Soviet and Afghan bombers continued to fly high for fear of Stinger.

Despite the impending Soviet withdrawal, the Soviet Air Force was given no rest, as Borovik records:

The Bagram aviation division worked days and nights. It dropped nearly two hundred tons of bombs every twenty-four hours, sometimes even more. For instance, during the 1988 operation code-named Magistral, which ousted armed rebels from the strategic road to Khost [near the Pakistani border] and delivered food and materiel to the blockaded town, the daily expenditure of explosives hit four hundred tons.

The Soviets were accused of conducting a scorched earth policy with aircraft and artillery to cover their withdrawal in early 1989. Certainly a slackening of the Mujahideen's SAM defences at Kandahar may have prompted the VVS and AAF to intensify their attacks. AAF MiG-21s armed with cluster bombs, defending Jalalabad, in some cases attacked at low altitude, improving accuracy. In the first two weeks of February the Soviet air force conducted 350 sorties over Afghanistan, helping safeguard the pull-out.

Despite the less than systematic air war, the Soviets learned the importance of combining helicopters with fighter-bombers in joint ground attack operations. The value of heli-borne infantry was fully appreciated and the Mi-24 proved a very good substitute for fighter-bombers in a CAS role. Moscow, though, had no answer to Stinger.

18

STORM IN THE DESERT

Overwhelming air power paved the way for a very swift ground war in 1991. It also featured top secret American cutting edge stealth technology. The tempo of the air campaign against Iraq during Operation Desert Storm was astonishing. A total of 109,500 combat sorties were flown, dropping 88,500 tons of munitions. The Coalition regularly flew more than 3,000 sorties every twenty-four-hour period, easily outstripping those rates achieved in Korea and Vietnam. The RAF flew more than 4,000 sorties, dropping 3,000 tonnes of munitions (including more than 100 JP233 airfield denial weapons, some 6,000 bombs, more than 100 anti-radar missiles and 700 air-to-ground rockets). The Iraqi Air Force was simply shot out of the sky, destroyed on the ground or fled.

It was estimated that Saddam Hussein's forces numbered about 200,000 when the ground war was launched; the air war was initially estimated to have accounted for half that number! When the fighting was over the air war death toll was revised to 10,000. The Iraqis themselves claimed they had lost 20,000 dead and 60,000 wounded in twenty-six days of sustained air attack.

Two powerful Republican Guard Corps armoured divisions invaded Kuwait on 2 August 1990. The following day a third moved to secure Kuwait's border with Saudi Arabia, sealing the country from the outside world. The small Kuwaiti Air Force briefly attacked the Iraqi armoured columns but its base was quickly overrun. Kuwait found itself under Saddam's control within just twelve hours. In response to the Kuwaiti crisis, the US instigated Operation Desert Shield, a massive multinational effort to defend Saudi Arabia in the event of Saddam advancing further south. The Coalition gathered 500,000 men from thirty-one countries.

Just five days after the invasion of Kuwait, American McDonnell Douglas F-15C/D Eagle multi-role fighters (from the US Air Force's 1st Tactical Fighter Wing) flew non-stop from Langley Air Force Base, Virginia, to Dhahran in Saudi Arabia. This feat took up to seventeen hours and required a dozen in-flight refuellings. In the following days, more than twenty American squadrons deployed to Saudi Arabia, including the first of twenty-two state of the art Lockheed F-117A Nighthawk stealth attack aircraft of USAF's 415th Tactical Fighter Squadron, 37th Tactical Fighter Wing.

The sinister-looking black Nighthawk with its distinctive angular lines was the first operational aircraft designed to use low-observable stealth technology. Its role was to penetrate densely defended areas, avoiding enemy radar and attacking high-value targets with pinpoint accuracy. By August 1990 the Lockheed Advanced Development

Company had built fifty-nine such aircraft at great expense to the US tax payer. The Americans were later to build the B-2 stealth bomber but few other countries could afford such advanced technology.

British RAF Tornado F3 interceptors and Jaguar GR.1A attack aircraft also deployed to Saudi Arabia. From the waters of the Gulf, the US Navy and Marine Corps provided the Grumman F-14 Tomcat, the McDonnell Douglas F/A-18 Hornet, Vought A-7E Corsair II and AV-8B Harrier II, to name but a few.

On the face of it the Coalition air forces had their work cut out. As a result of the Iran–Iraq War, Baghdad had invested heavily in air defence, creating a reasonably integrated system. Iraq's Air Defence Command was considerable, with approximately 7,000 surface-to-air missiles (SAM) and 6,000 anti-aircraft artillery (AAA) pieces directed by a national radar network. Similarly, the Republican Guard had its own system to defend key sites such as Baghdad, with sixty SAM batteries and 3,000 AAA guns.

At the outbreak of the war the Iraqi Air Force was allegedly the sixth largest in the world, with up to 750 fighters and ground attack aircraft in its bristling inventory. In reality, though, probably fewer than ninety of the air force's 113 Mirage F1s and only half its fifty MiG-29s were operable, and only about 100 of its 200 MiG-21s/F-7s remained. On top of this, the Iraqi Army Air Corps fielded about 150 combat helicopters.

The Coalition's combined air forces, deploying 1,800 aircraft, launched Desert Storm around the clock from mid-January 1991, targeting Iraqi command and control sites, Scud ballistic missile installations and lines of communication. Vitally, air superiority over Kuwait and

Iraq was achieved within twenty-four hours of the first air attacks. Many of the aircraft, including the British Tornado and American F-117A stealth fighter, had never seen combat before.

Flight Lieutenant John Peters, 15 Squadron RAF, piloting a Tornado armed with eight 1,000lb bombs and two sidewinder missiles, recalled:

Heavy Triple-A starts coming up, lazy curving arcs of tracer, dozens of bright points, streaming red droplets like a giant showerhead spraying skywards. The buggers are shooting at us! The shells burst into blossoms of black and white smoke, chucking out shrapnel in every direction. As well as being terrifying to look at … with our helmets on, we can hear nothing at all … It is like watching a silent film. The explosions are peppering up in continuous streams …

Shortly after, Peters and his navigator Flight Lieutenant John Nicol were hit by an SAM. The pair successfully ejected from their stricken aircraft and were captured.

The Iraqi Air Force lost almost 50 per cent of its fleet: 141 aircraft were claimed destroyed on the ground, 35 in air-to-air engagements, whilst 122 fled to Iran never to be returned. The national air defence system was quickly neutralised and the radar sites flattened. Three Iraqi aircraft attempted an Exocet missile attack on Coalition shipping on 24 January 1991. An intercepting Saudi F-15 shot down two and from that point the Iraqi Air Force took little part in the war.

Over a six-week period the Coalition air forces also focused on the Republican Guard's tanks and Armoured

Personnel Carriers (APC)s. Approximately 35,000 sorties were launched against Iraqi ground forces, of which 5,600 were directed at the Guard dug in over a 4,000 square mile area. Initial estimates of Iraqi losses to air strikes were about forty Republican Guard tanks, whilst the army lost fifty-two tanks, fifty-five artillery pieces and 178 trucks. General Schwarzkopf wanted a 50 per cent degradation of Iraqi fighting capabilities by the air campaign before committing himself to the ground war. Assuming that the estimates of Saddam's forces in the Kuwait theatre of operations (KTO) were correct, they more than achieved this. Just before the ground offensive the Coalition was claiming the destruction of up to 1,300 tanks, 800 armoured personnel carriers and 1,100 artillery pieces. Instead of withdrawing, Baghdad announced it would fight the 'Mother of battles'.

While the Iraqis claimed up to 200 Coalition aircraft, in reality twenty-seven US aircraft were lost in combat (including eleven to AAA and one to an SAM) and nine non-US aircraft were lost. Additionally, ten American and four allied planes were lost to non-combatant causes. A total of six RAF Tornados were to be destroyed in combat, of which two are thought to have flown into the ground, one was downed by its own bomb, and the rest were hit by SAMs.

Saddam decided to pre-empt the ground war, possibly in the hope of provoking the Coalition before it was ready. The action took place around the Saudi town of Al Khafji. Elements of the Iraqi Army's 5th Mechanised Division moved out over a 50-mile front on the night of 29 January 1991. Saddam was swiftly thwarted by Coalition air power.

During the early hours, a flight of Cobra helicopter gunships from 369th Gunfighter Wing hunted down the Iraqi armour with night vision goggles. They ran low on fuel and had to return to base, being replaced by four more Cobra, which destroyed a platoon-sized mechanised force. By 0500–0600 the Saudis realised that Saddam really meant business, as about twelve Iraqi armoured vehicles were observed on the western edge of the town. Nevertheless, no Iraqi tanks entered Khafji; they were all knocked out to the north, though some armoured fighting vehicles did get in.

Initially, Saudi commander Prince Khalid Bin Sultan al-Saud panicked, because it had been his decision to remove the Khafji garrison. It also seems he may not have informed his uncle, King Fahd, who understandably was furious that Saddam was now occupying part of his kingdom. General H. Norman Schwarzkopf, the Coalition commander, was horrified when he was informed that King Fahd wanted Khafji flattened by American bombers. 'I am sorry,' responded Schwarzkopf, 'we don't conduct ourselves that way. Can you imagine how it would look in the eyes of the world if the United States of America bombed a Saudi town into rubble just because a few Iraqis were there?'

The following day, under attack from A-10s, the other two attacking Iraqi battalions failed to get through. Inside Khafji, although a brigade of 2,000 had been thrown into the attack, the Iraqis only numbered about 600 defenders with infantry support weapons and some tanks on the outskirts. This solitary battalion, acting as Saddam's fist in Saudi Arabia, was to bravely hold out for two days. Although the town had no real military value, the

Coalition could not leave them, as there would have been a loss of face, plus a US Marine reconnaissance party was trapped in Khafji along with the Iraqi soldiers.

Just outside Khafji, Prince Khalid was frightened that the Iraqis might counter-attack and cut him off. He called up the air operations room in Riyadh and, after speaking to his own air force commander, was handed over to General Chuck Horner of the United States Air Force. 'I am worried,' confessed Khalid. 'We'll keep them off you,' Horner promised. He then added with a laugh, 'Khalid, I want you to keep one thing in mind.' 'What's that?' responded the prince. 'You're in a bunker in Khafji, and I'm here in Riyadh. It's easy for me to be calm!' The entire Iraqi garrison was subsequently destroyed.

Operation Desert Sabre was unleashed at 0400 hours on 24 February 1991. At 1515 hours the following day, spearheading the British 1st Armoured Division, 7th Armoured Brigade began to advance into Iraq, passing through the 1st US Infantry Division. They were to thrust eastward into Kuwait. Facing them were elements of the Iraqi 12th Armoured Division, which was now believed to be about 65 per cent combat effective; of its 250 tanks only 115 were operational. However, it was anticipated that the 12th and 48th Iraqi Divisions would remain in place, supported by an unidentified Iraqi brigade.

American Apache attack helicopters and A-10 Thunderbolt tank busters also played a significant role. One Apache alone destroyed eight T-72 tanks and on 25 February two USAF A-10 destroyed twenty-three Iraqi tanks, including some T-72s, in three close air support missions. Despite a ceasefire, the US 24th Division fought elements of the Hammurabi division again on

2 March after reports that a battalion of T-72 main battle tanks (MBTs) was moving northward toward it in an effort to escape. The survivors were forced back into the 'Basra Pocket'. By this stage Iraq only had about 700 tanks and 1,000 APCs left in the KTO and the fight was over.

After a ground war of just four days, the Iraqi Army of occupation had been effectively cut to pieces (no more than seven of the original forty-three Iraqi divisions were considered operational) and 15 per cent of Iraq was under Coalition control. In particular, its powerful armoured forces had not proved to be the formidable threat portrayed by the West. This was in part due to the Coalition's crushing air power. Whilst they were inferior in many technological aspects, the Iraqi armoured forces' poor combat performance must also be put down to inadequate leadership and strategy.

What happened to Iraq's 500,000 troops in the KTO? Having spent six weeks pinned down by Desert Storm's relentless air campaign, Iraqi morale was rock bottom and desertion rife. The Iraqi Army knew that Baghdad had abandoned it. Two Iraqi divisional commanders informed their British captors that they had received no orders for almost two weeks. Washington assessed that at least 150,000 Iraqi troops had deserted before Desert Sabre even commenced.

Washington estimated that more than 100,000 were killed and 300,000 wounded, and another 175,000 were taken prisoners of war. However, 575,000 is far more than was ever originally assessed to be in the KTO. British estimates were much more conservative, with 30,000 dead and 100,000 wounded. The air war alone was initially thought to have accounted for 100,000, but when the fighting was over it was revised to 10,000. It has also since been estimated that just 10,000 Iraqis were killed during the land offensive.

Despite the media reporting a 'turkey shoot' by Coalition fighter-bombers, most of the vehicles destroyed on the 'highway of death' or Highway 6 north of Kuwait City were empty. Analysis of photographs of Highway 6/ Muttla Pass showed that the bulk of the vehicles caught on the road were in fact civilian cars, minibuses, pick-up trucks, tanker lorries and even a fire engine taken by flee-ing Iraqi soldiers. There were very few military vehicles (including several armoured cars, lorries, fuel trucks and a tank transporter) actually on the highway. Such details mat-tered little following the successful liberation of Kuwait. When Desert Storm came to a close in early March 1991, Iraq only had about 150 of its fighters remaining.

19

NO SHOW

American troops triumphantly burst into Saddam International Airport and Rashid military airbase, both just outside Baghdad, in early April 2003. Only at the former was there any real resistance and, to the Americans' amazement, there was no sign of the once powerful Iraqi Air Force. During Operation Iraqi Freedom, as in Desert Storm in 1991, the Iraqis decided discretion was the better part of valour. Little was expected of their air force in 2003 – after all it had only put up token resistance twelve years earlier – however no one expected it to vanish completely.

After years of military sanctions it was assessed by Western intelligence that the Iraqi Air Force might still have 130 attack aircraft and 180 fighters. Of these, only 100 were deemed to be operational, enough to thwart any internal unrest but not the powerful United States Air Force. The Iraqi Air Force was probably the most educated of the Iraqi armed forces, with a greater appreciation of the impossible task facing it in 2003. The Americans had made a point of showing that they dominated Iraqi airspace. Iraqi pilots knew it was an encounter they could not expect to survive and did not intervene at all. The Iraqi Air Defence Command did what it could, but was simply overwhelmed and Regular Army efforts were at best half-hearted.

Intelligence chief General al-Samurri tried to warn Saddam but was sacked. One of his successors, former intelligence chief Farouk Hijazi, captured on 24 April 2003, had a similar tale to tell; Saddam would simply not accept that they could not successfully resist the technological array of weapons facing them when the US-led Coalition conducted Operation Iraqi Freedom. By this stage the loyalty of many of Saddam's generals had reached breaking point.

Along with the elite Republican Guard, elements in the air force knew they could not avoid the war altogether, but sought a way to safeguard themselves and their remaining airworthy fighters. It appears the Americans agreed that if they did not fight then al-Asad, about 170km north-west of Baghdad and home of the air force's Fighter Command and the second largest base in Iraq, would be spared. Secretly, across Iraq the order went out not to resist.

Saddam Hussein knew he faced betrayal. For example, former air force General Ali Hussein Habib was arrested just before the Coalition air attacks commenced on Baghdad. His headless body was located outside Abu Ghraib prison in a shallow grave on 15 April 2003. Habib had been involved with the Iraqi Chemical Weapons programme and was prepared to be interviewed by UN inspectors without minders. It may have been that Saddam's regime suspected he was already collaborating.

On 16 March 2003 at an Iraqi council of war in Baghdad, 150 senior officers including air force General Kareem

Saadoun dared not remind Saddam they simply could not win. In 1991 pilot training was poor, as was the service-ability of its fighters. On top of this, it was operating some fifteen different types of fixed-wing aircraft. The Iraqi Air Force was not blind to the fact that these deficiencies had only got worse. The UN embargo ensured they received no vital spares, no new aircraft or surface-to-air missiles, though some spares for Iraqi MiG-23s and MiG-25s may have been sneaked in via Syria.

Coalition intelligence on Iraqi dispersal airfields was first class, and the Iraqi Air Force was only too aware of the danger from Coalition Special Forces, which were ranging far and wide in their search for Saddam and his weapons of mass destruction. One solution to this was deception. The Iraqis had a lot of inoperable airframes and many of these derelict platforms were parked out in revetments as decoys. The challenge for the Coalition was to detect the ones that were operational. Saddam lost more than 100 aircraft to Coalition military action in 1991; this time round the battle damage assessment was much more difficult because of the number of Iraqi aircraft that were already little more than junk.

The Iraqis had a fleet of about ten Mi-24 Hind attack helicopters, sufficient for operations against the Kurds and possible insurrections but little else. Their Gazelle and Bo-105 helicopters were in a similar state. Likewise, it is doubtful that 100 of its transport helicopters such as the Mi-8 were airworthy. In the face of concerted Coalition attacks in early 2003, hiding them away became a priority.

Iraq's airbases and hardened aircraft shelters (HAS) were systematically targeted, as were Iraq's air defence, com-mand and control, and intelligence facilities. In the face of up to 1,400 Coalition sorties a day, none of Iraq's armed forces showed much initiative. Nine Iraqi SAM sites were attacked in downtown Baghdad on 26 March 2003 and on 1 April the US Department of Defense showed an F-15 delivering a precision-guided munition on a suspected SA-2 missile site south-west of Karbala. Missile maintenance facilities were also hit, such as the facility in Mosul, believed to service all types of missiles.

Nonetheless, Iraqi air defences gave AH-64D Apache Longbow helicopter gunships of the US 11th Aviation a nasty surprise when they attacked elements of the Republican Guard's Medina Division. The Iraqis resorted to that old, tried and tested expedient of human intelligence. On 24 March an Iraqi major general in An-Najaf was able to report the location of the Apaches' assembly areas, and the fact they were on the move, to Iraqi air defence personnel using a mobile phone. As a result, the Apaches came under intense ground fire and lost a helicopter. This success was short-lived for the area was soon subject to intense attack by American A-10 Thunderbolts and British Harrier GR7s.

Some of the Iraqi Air Force's potential dispersal sites were seized within the first forty-eight hours of the war. British and Australian SAS Special Forces were used to secure the air bases known as H–2 and H–3 in Iraq's western desert. Iraqi aircraft dispersed on H-2's airfield were destroyed on the ground by American AC-130 gunships called in by the Special Forces.

From H–2 Special Forces were able to Scud missile hunt and direct aircraft, principally A-10 tank busters, against ground targets to the west and south of the airfield.

Similarly, air defence radar in the western desert near H-3 was targeted by Coalition bombers. Mudaysis airfield, also in western Iraq, was targeted when an F-16 put a precision-guided munition on to its radar site. By late March the Coalition was staging air operations from a number of Iraqi airfields under its control. Tallil, outside An Nasiriyah, was the first forward air base from which the Coalition aircraft were able to operate.

British operations in the south quickly neutralised Az Zubayr airbase near Basra. The American thrust north ensured that Karbala North-east (a civilian site) and the military facilities at Habbaniyah and Baghdad Muthenna were non-operational. By 1 May the Habbaniyah area was occupied by the US 3rd Armored Cavalry Regiment.

When US Marines overwhelmed the Special Republican Guard and militia forces defending Saddam International Airport, just 12 miles to the west of Baghdad, on 3–4 April 2003 they shot up the civilian aircraft, mostly already laying wrecked on the runway. The facility was both military and civilian and the Marines made themselves at home in the airport's HAS, which contained nothing but thin air. The US 58th Aviation Regiment soon had the control tower up and running and the airport was symbolically renamed Baghdad International.

To the south of the capital, US Marines, having destroyed the Baghdad Republican Guard Division, seized An Numaniyah airbase. To the east of Baghdad, American forces expected to find the vast Rashid airbase stiffly defended by Iraqi Air Force personnel, Special Republican Guard and Republican Guard. Notably, Rashid was home to the Republican Guard Squadron and the Special

Transport Squadron. Instead, they entered the facility without facing any real resistance. Of the air force there was no sign; once again they had fled.

US Marines secured al Amarah airbase and interceptor operations centre on 8 April. Three days later American forces entered Tikrit, securing Al Bakr airbase (home to MiG-23s, Su-24s, Su-22s and various transport squadrons) and Tikrit East and South. The brief war with the Iraqi Air Force was all but over.

However, resistance continued on some Iraqi airbases. In mid April the newly arrived US 4th Infantry Division fought a brief firefight near Al Taji Airfield north of Baghdad, capturing amongst other things an SAM warehouse. By 16 April the Americans also had control of Samarra, thereby securing the military airbase at Samarra East.

The Coalition found Iraqi air defences cynically placed in civilian sites to deter air attack. For example, in Baghdad an Iraqi artillery unit, with ammunition placed at various points, was deployed in a park in the middle of a residential neighbourhood across the street from a girls' school. It was apparent that the Iraqis would fire a gun, move it, fire again, and continue the process as a way of avoiding being bombed. In Al Kut more than twenty anti-aircraft guns were removed from an amusement park by US Marines.

Once the fighting started, according to some Iraqi Air Force officers they received no further orders. Colonel Diar Abed, at Rashid airbase, noted, 'We had no orders. We just stayed in the bases and waited … Why don't they give us orders. The leaders at the base didn't know anything.' General Saadoun, also at Rashid, recalls bitterly, 'They just gave us Kalashnikovs [assault rifles], not even anti-aircraft weapons.' Two weeks before Rashid fell, its communications were cut. Somebody, somewhere had betrayed them. The Coalition Blitzkrieg sliced through Iraqi defences and rolled into Basra and Baghdad with relative ease.

Evidence indicates that the Iraqi Air Force was either bought off like the Republican Guard or simply threatened into submission. US military planners wanted to secure Saddam International Airport and take out the main Iraqi fighter bases at al-Asad, al-Taqqadum and Rashid, so they cut a deal with some elements of the air force. It remains unclear what level of complicity air force commander Lieutenant General Hamid Raja Shalah al-Tikriti had with Washington, but the fact remains the Iraqis did not put up a single aircraft to resist the Coalition.

Intriguingly, unlike the senior Republican Guard and intelligence officers who are believed to have betrayed Saddam Hussein, General Shalah was rated seventeenth in America's 'most wanted' deck of cards. Ultimately though, in the face of Operation Iraqi Freedom, Saddam's Air Force had simply fled the skies and avoided battle.

20

CALL SIGN UGLY

Apache attack helicopters flown in Afghanistan by the US, British and Dutch armies regularly engaged large numbers of Taliban fighters using their full range of deadly ordnance. The British Apache AH1 is the UK's AgustaWestland licence-built version of the American AH-64D Apache Longbow. In Afghanistan it operated under the call sign 'Ugly', followed by a numerical designation. 'On paper, the British Apache was the most expensive – and best – attack helicopter in aviation history,' said army helicopter pilot Ed Macy. 'As the most technically advanced helicopter in the world, the Apache AH1 was also the hardest to fly.'

Two British Army Air Corps regiments (each with three squadrons) operated the Apache, which first deployed to Afghanistan on combat operations in 2006 in support of Task Force Helmand. They also deployed over Libya in 2011. Prior to the British withdrawal from Helmand, the 3rd Regiment AAC operated its Apaches in Afghanistan from January 2011 to January 2012, with its three squadrons each deploying on a four-month tour. No. 662 Squadron served first, then 663 followed by 653 Squadron, which undertook its official handover on 20 September 2011. However, 653 did commence operations the week before.

One of the most dramatic incidents involving the Apache occurred on 13 January 2007. Ugly 50 and 51 from 656 Squadron recovered the body of Lance Corporal Mathew Ford of 45 Commando from Jurgoom fort, a Taliban stronghold. None of the rescuers were wounded during the mission and they later received gallantry awards, including Ed Macy, who was awarded the Military Cross.

Despite their devastating firepower, British Uglies did not have it all their own way in Afghanistan – far from it. Taliban rocket-propelled grenades and heavy machine guns posed the greatest threat to helicopters. Apaches regularly had holes punched in them following combat missions. As well as small arms including assault rifles and light machine guns, the Taliban had a range of heavier weapons that also posed a potential threat to aircraft. These included the old Soviet-designed SA-7 surface-to-air missile, with a range of 4,000m, and the Soviet-designed 14.5mm and 12.7mm heavy machine guns, which have effective ranges of 3,000m and 2,000m respectively. Typically, British Apaches engaged targets at 2,000m.

The Taliban were also judged to have some of the newer SA-14s and Chinese HN15s, and even the odd US Stingers and British Blowpipe surface-to-air missiles, but few if any were operable. 'Working SAMs are a highly prized commodity to the Taliban, an Intelligence Corp briefer told us in 2007,' recalled Ed Macy. 'We reckon that the few they retain will be used only as a last ditch defence for very senior people; they will only be fired if a Taliban or al-Qaeda leader's life is under imminent threat.'

Macy and his colleagues were rattled when the Taliban unsuccessfully attempted to shoot down a Dutch F-16 jet fighter using an SA-7. He observed:

> Some days before the launch, a radio intercept heard a Taliban commander saying, 'Fetch the rakes and spades to hit the helicopters.' Initially the intelligence cell had assessed rakes and spades meant Chinese rockets and a launcher, we now had to assume they meant SAMs.

They were also alarmed when intelligence indicated the Taliban were planning to move a Stinger missile to Sangin or Kajaki to use against British helicopters. However, fortunately the Taliban had no way of replacing the Stingers' depleted batteries, which dated from the late 1980s.

While perhaps not foolproof, Apache crew were protected by the Helicopter Integrated Defensive Aid System, or HIDAS, which was constructed to defeat all known SAMs and did it automatically. The system detects every missile threat or laser or radar trying to lock in on the helicopter and automatically launches the necessary countermeasures such as chaff to confuse radar-tracking SAMs or flares to confuse a heat-seeker. If the threat was manually laser-guided then the warning system gave evasive manoeuvre commands. There was only one high-profile attempt to shoot down an aircraft in Afghanistan using an SAM, but the danger remained continuous nonetheless.

The British armed forces are governed by very strict rules of engagement, which dictate when they are permitted to open fire, either when in direct contact with the

enemy or to prevent imminent attack. This is designed
to minimise collateral damage. Details of these rules of
engagement are not made public for operational security
reasons. Taliban fighters could go quickly from insurgent
to civilian by simply discarding their weapon – giving
them more information on when they can and cannot be
fired upon simply hampered operations. This meant British
Apache pilots often had to make very difficult split-second
decisions when choosing to engage.

The typical ordnance fit on this devastating piece of
equipment is the 30mm M230 chain gun, Hydra 70/CRV7
rockets and Hellfire air-to-surface missiles – the human
body cannot withstand a direct impact from any of these.
The Longbow Apache can operate in all weathers, day or
night, and incredibly detect, classify and prioritise up to
256 potential targets in a matter of seconds. This makes it a
truly deadly weapon system.

The attack helicopter has evolved since the days of
the Cold War into a key asset when it comes to a theatre
commander's options in using rotary, fixed-wing aircraft
or drones to provide vital close air support. The Boeing
AH-64 Apache attack helicopter was first introduced into
US Army service in 1986. The AH-64D, with the highly
distinctive bulbous Longbow radar above the rotor blades,
appeared in 1997. The Apache has since seen service in
Afghanistan, Kosovo, Panama and Iraq. It is in service with
more than half a dozen countries, including the UK.

The Apache's array of ordnance is deployed using
the Target Acquisition and Designation Sight (TADS)
and the Pilot's Night Vision System (PNVS). Targets
are detected using the TADS or the fire control radar.
TADS gives pinpoint accuracy if time allows; otherwise

the helmet-sighting system can be used as this gives increased spread.

It is almost impossible to hide from the Apache's forward-looking infrared (FLIR) sensors. Movement and body heat are a giveaway, so playing dead, scattering or trying to hide offers no salvation. An Apache can loiter over a target area until its ordnance is completely expended. Essentially, Taliban fighters had nowhere to hide.

Apache gun camera footage from Afghanistan shot through the FLIR showed an effect similar to mortar or light artillery fire when the M230 chain gun was fired and appeared indiscriminate due to the overlapping and sizeable explosions around the target/targets. The effect is devastating. Individuals in the fire zone quite often kept running even when peppered with cannon fire – the kinetic energy given off by the rounds gives the targets an added impetus but does not mean they remain unscathed; it causes severe tissue trauma.

The formidable American Hughes M230 electrically operated Chain Gun is a 30mm single barrel automatic cannon mounted underneath the fuselage. This is the AH-64 Apache's standard armament Area Weapon System. Notably, 30mm cannon fire remains the most accurate air support option in the military's arsenal, and also delivers the lowest amount of collateral damage. The gun is normally fired in very short bursts (with ten rounds per second in pre-selected bursts of ten, twenty or fifty rounds), so it is easy to monitor the number of rounds put onto a target at any given time.

The Lightweight 30mm M789 High Explosive Dual Purpose (HEDP) ammunition is the primary tactical round, though the M799 High Explosive Incendiary (HEI)

round is also available. The casing is made from an alloy
rather than brass or steel to keep the weight to a minimum.
The Apache can carry up to 1,200 rounds. The M789 is
designed to provide suppressive firepower at ranges similar
to those of Hellfire or tube-launched, optically-tracked,
wire-guided (TOW) missiles, making it ideal for deal-
ing with enemy armour and providing wide area lethality
against enemy personnel.

During typical fire missions in Afghanistan, British
AAC Uglies flying in support of 3 Commando Brigade
in 2008–09 expended bursts of 20, 40, 125 and 400
30mm rounds that resulted in anything from four to
twelve fatalities. The UK spent $92 million in 2008–12
on a series of contracts to procure 30mm rounds from
the US manufacturer for its Apache fleet. In the second
half of 2011, British Apaches fired a staggering 55,000
30mm rounds in six months, averaging just under 10,000
a month. Thanks to operations in Afghanistan and Iraq,
during 2009–11 the ammunition manufacturer provided
the US Army with more than 700,000 30mm rounds.
In total it produced well over 10 million lightweight
30mm cartridges.

The M230 cannon can slip out of alignment and needs
to be harmonised, plus it has a tiny 3mm error. Although
the gun has a range of 4,000m, normal engagement range
is 2,000m – which means up to 6m error, other variables
can have an impact on accuracy as well. Nonetheless, the
gunner does not need to completely pinpoint his targets
with the sights, a task partially hampered by the movement
of the helicopter, because of the overlapping fragmentation of the M789 HEDP round, plus tracer rounds are
visible to 1,500m.

As well as the M230, the British Apache is also armed with Canadian Rocket Vehicle 7 (CRV7) rockets rather than the American Hydra 70, and uses High Explosive Incendiary Semi-Armour Piercing (HEISAP) or Flechette tungsten dart warheads. This is an unguided 2.75in (70mm) folding fin ground attack rocket. Typically an Apache can carry on each of its stub wings one M261 pod, each holding nineteen CRV7 rockets and four rail-mounted Hellfire missiles – giving a total of thirty-eight rockets and eight missiles. The CRV7 was chosen on the grounds it is faster, has a longer range and is 40 per cent more accurate than the American Hydra and has 95 per cent more kinetic energy. It has a maximum effective range of over 4,000m. High explosive or Flechette warheads were typically used against enemy fighters, with predicable results.

Ed Macy recalled:

Nothing beats a Flechette for multiple personnel out in the open. It was designed to burst open 860 metres into its flight, freeing its cargo of eighty five-inch long Tungsten darts. An explosive charge powered them onto the ground at speeds of Mach 2 – well over 2,640mph – shredding everything within a fifty metre spread. Each dart's intense supersonic speed created a huge vacuum behind it. If it hit a man in the chest, that vacuum would suck away everything in its path, and was powerful enough to tear flesh and muscle from a human target if it passed within four inches of one.

The AGM-114 Hellfire is an air-to-surface missile originally designed in the 1980s for an anti-armour role. There are a large number of variants with a range of up to 8,000m.

Laser and radar guidance for the Hellfire II and Longbow Hellfire is provided by nose-mounted opto-electronics on the AH-64 and the Longbow radar respectively. The Apache can carry up to sixteen such missiles if the rocket pods are not fitted. Hellfire missiles come with high-explosive anti-tank and blast fragmentation warheads. The AGM-114M/6 Hellfire II was designed to deal with light vehicles, bunkers, buildings and caves. It was used regularly in Afghanistan to neutralise Taliban strongpoints.

Elements of 3 Regiment, AAC, served in Afghanistan from January 2011 to January 2012 as part of Task Force Jaguar and flew in support of Operation Herrick 14. During this time Apache pilot Captain Antony Thompson was awarded a commendation by the commanding general of the US Marine Corps 2nd Aircraft Wing for his skills and professionalism. In addition, 662 Squadron was reported to have the highest 'kill rate' of any unit serving in Afghanistan. The Uglies finally withdrew from Afghanistan along with the rest of Britain's Task Force Helmand at the end of 2014. Because of what they did in Afghanistan, the Ministry of Defence gave Apache crews the same protection as Special Forces. Analysis conducted for the US Marine Corps found that 'the Taliban tended to break contact when air support arrived – not because air strikes inflicted heavy casualties, but because air attacks made it too dangerous to continue fighting'.

21

BOMBING TRIPOLI

Air power was used to help facilitate regime change once more in 2011. The international community was heavily criticised for its complete lack of urgency in helping the Libyan rising against Colonel Gaddafi. In reality it was quietly and efficiently planning to stop him from crushing the rebels. Judging by events on 20 March 2011, preparations were well under way before UN Security Resolution 1973 was ever agreed. The latter stipulated that a no-fly zone should protect civilians, thereby authorising NATO airstrikes against Libyan military targets.

Following the Libyan rising in Benghazi in February 2011, the rebellion against Gaddafi's rule quickly spread. He knew he needed to swiftly regain control of the border with Tunisia to cut the opposition off in the west of the country. By 11 March it was claimed he had retaken both Zawiya and Ras Lanuf. Once they were secured he could draw a firm line in the sand at the latter town and then, depending on international reaction, commence a vigorous campaign of roll back eastwards towards Benghazi. In preparation for this the Libyan Air Force began conducting increasing air raids in an effort to destroy weapons dumps in opposition hands in the Benghazi area. Inevitably, civilians were caught in the crossfire.

Without greater assistance from deserting members of the Libyan armed forces the opposition's options remained very limited. Small arms, rocket-propelled grenades, anti-aircraft guns and 'technicals' (pick-up trucks fitted with weapons) were no match for tanks, heavy artillery, rocket launchers and bombs. Opposition fighters were seen with SA-7s, but it was unclear if they were operational and it is not a user-friendly weapon.

While Gaddafi initially relied on tribal levies and mercenaries for his security operations, there were increasing indications that the bulk of the armed forces had remained loyal. Tanks were paraded in the streets of Tripoli in a show of strength. These were believed to belong to the Khamis 32nd Brigade, which could muster a single tank and mechanised regiment for offensive operations.

The Libyan armed forces inventory included a plethora of largely ageing Soviet-era weapons. Nonetheless, in semi-capable hands against ill-equipped and ill-trained rebel forces these were more than up to the job. On paper Gaddafi had almost 3,000 tanks, armoured infantry fighting vehicles and self-propelled guns, essentially enough for ten armoured divisions, but only the manpower to operate a fraction of these. His most modern tank was the ageing T-72, but courage alone would not stave off determined attacks by these 45-ton monsters.

Gaddafi's air force had a considerable array of jet fighters, numbering more than 400 combat aircraft, but few competent pilots to fly them. Gaddafi, though, had little need of his fighters, his key assets were the air force's ground attack aircraft, which included MiG-23s, Mirage F-1ADs and Su-22s, and his Mi-24 Hind helicopter gunships. The Su-22 had been used to attack rebels at Ajdabiya and Ras Lanuf. Such bombing raids were initially inaccurate and half-hearted, but grew in intensity. It was quite probable mercenaries conducted some of these raids; Zimbabwe was believed to have provided a few pilots, who acted as a force multiplier. On Libya's exposed coastal roads rebel convoys remained very vulnerable to air attack.

The blame for Gaddafi's well-stocked arsenal used against the opposition could not be entirely laid at the feet of the former Soviet Union. Gaddafi's war machine was totally reliant on foreign manufacturers for spares and ammunition. Unlike neighbouring Egypt, Gaddafi's Libya had no real military manufacturing capability. Lacking the skilled labour to develop an indigenous arms manufacturing sector, he was content to import all his needs over the years.

Up until the mid-1960s Libyan arms imports were extremely small, but with the boom in oil revenues that swiftly changed. The major expansion of the regular army coincided with Colonel Gaddafi's coup d'état, which swiftly realigned the country from its traditional relationship with America, Britain and France to the Soviet Union and Egypt. American and British military bases were closed down and western military aid ceased. It was at this point that Soviet-supplied arms including tanks, artillery, fighter jets and patrol boats began to flood into the country. Since then, however, Gaddafi had also bought weapons from a host of other countries.

As well as the aircraft supplied by the Soviet Union, Gaddafi's relationship with France flourished, culminating in a deal for more than 100 Mirage IIIs and 5s in the mid-1970s worth £110 million. The last of his French-supplied Mirage F1 jets were delivered in the late 1980s; Iraq, the main export country, found them effective in an anti-shipping role but they were outmatched during the 1991 Gulf War. A decade earlier, in 1981, two Libyan Su-22s had been shot down by US Navy Tomcats in the Gulf of Sirte air battle. By the 1990s Gaddafi had purchased more arms than he could possibly need and deliveries had begun to dry up.

By this stage Gaddafi possessed one of the largest and most potent air forces in North Africa, with a long-range capability provided by half a dozen Tu-22 supersonic bombers. However, these aircraft were expensive and difficult to operate, and it was highly doubtful that they were operational. The bulk of Gaddafi's aircraft were the Soviet-supplied MiG-23, which was robust and easy to produce in vast numbers. As a fighter, though, the MiG-23 was uninspiring; it avionics and performance were found wanting up against its American counterparts and its cockpit offered poor visibility. Nonetheless, Gaddafi had a number of the MiG-23BN ground attack variant known as the 'duck-nose', which were useful against the Libyan opposition.

NATO's opening air attacks on Libya on 20 March 2011 were not reminiscent of the 'shock and awe' campaign against Iraq in 2003, as some might have expected, as they were nowhere near the same intensity. Nonetheless, the speed of the assault came as a surprise to everyone. Within an hour of French fighters attacking Gaddafi's tanks south-west of beleaguered Benghazi, the Coalition sought to 'shake the battle space' with missile strikes on more than twenty Libyan air defence sites in western Libya. This was designed to secure battle space dominance before combat patrols enforced a no-fly zone. French fighters were operating from French airbases at Istres and Solenzara (with ninety- and sixty-minute flying times respectively).

At 1500 Eastern Time the first of 112 Tomahawk cruise missiles hit their targets. A US spokesman confirmed that the US Navy was using a new generation of Tomahawk that could loiter over the target before striking – although on this occasion they were used in the conventional manner. The missiles were fired from US destroyers and submarines that constituted part of a fleet of twenty-five vessels gathered for Operation Odyssey Dawn. A British Trafalgar-class submarine also contributed to the bombardment.

The attacks were the opening phase of the Suppression of Enemy Air Defences as the Tomahawks were directed at Gaddafi's integrated air defence system, focusing on his SA-5 surface-to-air missiles and their command and control centres. Gaddafi's antiquated Soviet-supplied surface-to-air missiles were well past their sell-by date and many of them were considered non-operational.

His air defence command was assessed to have approximately thirty sites controlled by fifteen early warning radars. Missiles included SA-2, SA-3, SA-5, SA-6 and SA-13. These forces were organised into around fifteen brigades. The SA-6 in the hands of the Bosnian Serbs proved capable of shooting down a US F-16 fighter jet and the SA-6 offered a stand-off capability with a 300km range.

In the air the RAF's Tornados and Typhoons bore the brunt of things along with American F-16s based in Italy.

Canada also sent half a dozen CF-18 jet fighters (which last saw combat over Kosovo) and Belgium, Denmark, Greece and Norway each deployed half a dozen F-16 fighters to Sicily for immediate action. The American carrier USS *Enterprise* and her carrier strike group were in the Arabian Sea. *Enterprise* could be redeployed via Suez at short notice; her air wing comprised up to eighty jets.

In the meantime, an Amphibious Ready Group or Expeditionary Strike Group was on the way from America. This was headed by the USS *Bataan*, another assault carrier, whose Marine Corps air wing can include AV-8B Harriers and attack helicopters. These vessels are the largest amphibious assault ships in the world. *Bataan* was supported by the amphibious transport dock ship USS *Mesa Verde* and the dock landing ship USS *Whidbey Island*. They deployed from America on 23 March to join the US 6th Fleet based at Naples, Italy. The US Marines had ample experience of this type of operation, *Bataan*'s sister ship *Kearsage* saw action during the Balkans Conflict. Clearly, though, one assault carrier was not sufficient. During the Balkans crisis NATO maintained four carriers on station.

While the British frigates HMS *Cumberland* and HMS *Westminster* were off Libya, the UK's naval air force was few and far between after the withdrawal from service of the carrier HMS *Ark Royal*. HMS *Ocean*, the helicopter carrier, could deploy Apache attack helicopters, but the Albion Class assault ships were not designed to support no-fly zone ops. Meanwhile, the French carrier FS *Charles de Gaulle* based at Toulon was poised to play a role. She could contribute up to forty aircraft including the Rafale. Likewise, the Italian carrier *Giuseppe Garibaldi*, which carries AV-8B

Harriers, stationed at Taranto was available, although Italy only offered basing facilities.

It all seemed so promising when potentially overwhelming Western air power rode to the rescue of the beleaguered Libyan rebels. The opening bombardment was an impressive display of military muscle. However, following six weeks of the NATO-led Operation Unified Protector, conducted under UN Security Council Resolutions 1970 and 1973, Libya remained in a state of civil war.

While publicly this intervention was launched on humanitarian grounds, everyone knew that the unspoken goal was the removal of Gaddafi – yet two months on from the start of the rebellion he remained firmly ensconced in Tripoli despite some 7,000 sorties. Furthermore, fighting continued across Libya. In Benghazi the Libyan opposition continued to train its forces, but they were a long way off forcing Tripoli from Gaddafi's grasp.

There was a terrible sense of déjà vu – the Balkans and Iraq sprang to mind, where air power was expected to achieve swift and decisive political goals. Serbia and Kosovo proved otherwise: Serbian-backed forces clung on around Sarajevo, and although finally driven from Kosovo, Slobodan Milosevic's Serbian Army escaped largely unscathed. Meanwhile, in Iraq Saddam Hussein went about his daily business unhindered by the long-running no-fly zone. Although the high-tech airstrikes on Gaddafi's air defences and ground forces were impressive, his command and control structure remained largely intact.

Nevertheless, by July 2011 the rebels were 45 miles from the Libyan capital Tripoli, having made significant inroads to both the south and east of the city. Whether this was mainly due to an improvement in rebel military capabilities (in part thanks to French weapons supplies), NATO air attacks or falling morale amongst Gaddafi's supporters remained unclear. Ultimately it was a combination of all three as both sides continued to slog it out.

After gathering their strength in the Nafusa Mountains near the Tunisian border, the rebels attacked south-east toward Al-Qawalish from their strongholds at Kikla and Qala. As well as the usual assortment of armed pick-up truck 'technicals' for the Al-Qawalish attack, the rebels also fielded armour that included tanks, mechanised infantry combat vehicles and self-propelled anti-aircraft guns (the latter can lay down devastating fire on ground targets). In the meantime, to the north the rebels also pushed west from the city of Misrata and were at least 4 miles from their start line at Dafniya, putting them in control of the Na'imah area.

It was not until August that the rebels finally took Tripoli and ended Gaddafi's hold on power. He and his remaining supporters fled south but were caught by a NATO air strike that destroyed fourteen vehicles. A local militia captured Gaddafi and promptly killed him on the spot. The Western obsession of applying air power as surgically and cleanly as possible ignored the fact that achieving military goals ultimately boils down to taking and denying ground to the enemy. The application of air power alone is never enough. In the case of Gaddafi's once powerful armed forces, after years of neglect they simply dissolved as they were no longer prepared to support his rule.

22

MODERN COMBAT TRENDS

Since the end of the Second World War air power has continued to be seen as a war-winning weapon. Whether it be conventional war, low-intensity conflict or UN-sponsored policing actions, the utility of air power continues to be considered a highly successful tactical and strategic option. Indeed, in the closing decade of the last century the most dramatic and effective use of air power as an extension of the political means occurred over Bosnia and Serbia. There, NATO air forces were able to prevail against the Serb military with negligible losses. In the case of the Kosovo crisis, NATO was able to impose its will on Serbia with the loss of just one aircraft to military action.

Surprisingly, post-1945 just 30 per cent of all air combat losses have been as a result of air-to-air engagements, a ratio in fact similar to that established during 1939–45. Since then approximately 2,000 aircraft have been lost in dogfights compared to approximately 5,000 to all other causes.* Analysis of regional air combat casualties shows that this trend remains largely consistent across the board even today. In the Balkans, dogfights accounted for just 25 per cent of the losses. Only a few anomalies emerge. In the Middle East, the Israelis' outstanding victory over the Syrian Air Force in 1982 saw the trend reversed, with 70 per cent of losses incurred by the Syrians during air-to-air combat. In sharp contrast, African conflicts tended to be characterised by close air support (CAS) ops with almost no air-to-air warfare whatsoever.

According to the US Air Command and Staff College's 'Achieving and ensuring Air Dominance', during the Second World War air-to-air combat accounted for 32 per cent of losses in the European theatre of war and 41 per cent in the Pacific. Since then the ratio had varied from 30 to 63 per cent. This accords with extensive recent research looking at every major conflict since 1948. Analysis shows that approximately 60 per cent of all air-to-air combat losses occurred in the Asia–Pacific region, with the Middle East second at just over 30 per cent. As already noted, Africa bucks this trend with 90 per cent of military air casualties resulting from other causes (i.e. ground fire, accidents, etc.).

Major dogfight losses are now but a distant memory. The last significant air-to-air combat casualties occurred during the 1950–53 Korean War, followed by the 1963–72 Vietnam War and then the 1973 Yom Kippur War. The latter is most remarkable as the engagements occurred over a much shorter period than the two previous wars. Israel's

* It is impossible to be precise, due to differing counting procedures and contradictory reporting. What these figures do clearly illustrate is the fundamental trends in losses during the various conflicts.

victory over Syria in 1982 pales into insignificance when compared to these bloody conflicts.

Since the 1960s, time and time again surface-to-air missiles (SAMs), particularly man-portable air defence systems (MANPADS), have had a significant impact on the conduct and outcome of wars, such as the struggles in Afghanistan in the late 1980s, Angola in the early 1980s, Egypt in the early 1970s and Vietnam in the mid-1970s. In particular, Soviet exports of the SA-7 and the American Stinger to the world's trouble spots became prominent at the end of the 1980s.

Indeed, since the 1970s, with the exception of the 1982 Lebanon War, air war losses have consistently been caused by ground fire, air strikes or accidents. In the period under review approximately 2,900 military aircraft had been brought down by ground fire (AAA, SAMs and small arms). Coalition air losses during Operation Enduring Freedom launched against Afghanistan in 2001 were exceptionally low despite the ever-present threat of MANPADS. These included an American Chinook helicopter lost during Operation Anaconda in March 2002, brought down by a Taliban rocket-propelled grenade.

During the air war over the Democratic Republic of Congo (DRC, formerly Zaire) in 1998–2001, the number of aircraft brought down by ground fire is impossible to validate, but estimates suggest they included a Zimbabwean Air Force C-212 light transport, a Zimbabwean helicopter, a SF260 Warrior (possibly Congolese), two MiG-21s (possibly Angolan), two Namibian Air Force Alouette helicopters and two other unspecified types. Whilst the Ethiopia–Eritrea conflict of 1998–2000 saw some notable dogfights, the bulk of the Ethiopian aircraft lost were brought down by ground fire, including approximately fourteen MiG-21/23 fighters, one Su-25 fighter-bomber and six Mi-24 helicopter gunships.

In the 1999 Kosovan air campaign NATO remarkably lost just a single fighter to air defences, an American F-117 Nighthawk, despite Serbia's 1,800 AAA pieces and array of Soviet-designed SAMs. In the summer of 1995 NATO launched its military offensive against the Bosnian Serb forces waging 'ethnic cleansing' against the Bosnian Muslims. In eleven days Operation Deliberate Force, supported by Bosnian and Croat ground offensives, launched 3,515 sorties. Once again NATO lost one aircraft to ground fire: a French Mirage 2000K was shot down by an SA-7 MANPAD on 30 August 1995.

In Sri Lanka the Tamil Tiger guerrillas, lacking credible anti-aircraft systems, shot down only about nine aircraft (three fixed-wing – two SF-206TPs and one Cessna 337 Super Skymaster; and six helicopters – two bell 206s and four Bell 212s) in the six-year period 1983–89. They tended to get better results launching commando raids directly on Sri Lankan airbases.

In Latin America during the counter-insurgency conflicts of the mid 1980s, El Salvador and Nicaragua were able to conduct CAS ops against the Farabundo Martí National Liberation Front (FMLN) and Contra guerrillas with little threat to aircrews. In 1984 a Salvadoran Huey helicopter was allegedly shot down by machine gun fire, although the government claimed it crashed. The FMLN reportedly received at least one shipment of SA-7s in 1989, which helped hasten the peace process. However, there seem to have been few high-profile shoot-downs. The Nicaraguan Sandinista government lost one of its two Mi-8 helicopters, which were shot down by guerrillas in January 1983; using SA-7s, the Contras destroyed a second in 1985. Despite this, the Contras were unable to counter the Sandinistas' Mi-24 helicopter gunships. The following year it was the Contras who had a supply plane shot down, to the embarrassment of the CIA.

In the 1982 Falklands conflict, Britain lost two Sea Harriers and three Harrier GR.3s to ground fire, whilst British Rapier missiles claimed nine Argentine aircraft; Sea Dart, eight; Sea Wolf, five; and Sea Cat, six. During the Israeli invasion of Lebanon in 1982 the Israelis lost only three jets (an A-4, an F-4 (S) and a Kfir) and about four helicopters to ground fire. The Iran–Iraq War was characterised by a lack of air-to-air combat and most losses were due to air defences. When Iran retaliated against the Iraqi invasion by attacking Baghdad, sixty-seven aircraft were shot down. By October 1980, Iran had lost ninety to 140 combat aircraft. These losses along with manpower and material shortages, particularly of US-manufactured spares, stymied the Iranian Air Force from that point on. The Iranians claimed to have shot down thirty-four Iraqi jets during the invasion.

Nevertheless, the Iraqis suffered increased air losses when the Iranians were able to make qualitative improvements to their air defences. They succeeded in giving the Iraqi Air Force a particularly bloody nose at the beginning of 1987 using American-supplied Hawk SAMs. Iraq, having suffered such high losses in aircraft, also lost a large number of precious pilots who were even harder to replace.

African experiences of air war in Angola, Ethiopia, Mozambique and Sudan have been characterised by CAS operations, countered by the threat of SAMs and AAA.

The ubiquitous Soviet-designed MiG-23 was often in the forefront of the action. During the Ogaden War 1977–78 the Somalis lost thirteen aircraft to ground fire and the Ethiopians just three, leaving their air force free to support ground operations.

This lesson was not missed and the fortunes of the Angolan Civil War were swayed by the use of SAMs. The struggle became punctuated by a long list of shoot-downs. In 1980 the National Union for the Total Independence of Angola (UNITA) opposition forces, using Soviet SA-7s, shot down two Marxist government transport aircraft. The following year they accounted for a government transport plane. In 1983, UNITA, using a combination of SA-7 and AAA, claimed five MiG-21s and four helicopters.

By 1984 the Angolan government's growing airpower, with the arrival of twenty-five Soviet-supplied MiG-23 fighter-bombers and twelve Mi-24 attack helicopters, began to turn the tide against UNITA, despite the threat of SA-7s. South Africa's (UNITA's principal backer) Mirage F1s and IIIs were no match for the MiG-23s and SA-8s deployed by the Angolan government forces and its Cuban allies, and they kept well clear. America's response to the presence of these Soviet-supplied aircraft was to provide UNITA with Stinger SAMs. This helped combat the Angolan government's air power. Using this weapon, in 1987 UNITA shot down two MiGs, including a Cuban one on 28 October 1987.

Also that year, UNITA's use of SA-7s prevented the Angolan government from intercepting South African Air Force fighter-bombers supporting UNITA ground forces. The following year UNITA claimed two helicopters, two MiG-23s and two transport aircraft. By mid-1989 the annual tally was one helicopter and three MiG-21s and the following year five Mi-24s and three MiG-23s. Shortly after, Cuba and South Africa withdrew, forcing the Angolans to the fruitless negotiating table.

During the second Sudanese Civil War, Sudan's Air Force proved incapable of search and destroy missions against the Sudan People's Liberation Army (SPLA), to the extent that by 1985 it had not even made a positive sighting of the guerrillas. In contrast, in December 1983 the Sudanese Air Force lost three helicopters, at least one of which is believed to have been shot down by the SPLA, costing the Sudanese Air Force seven senior officers and its most experienced helicopter pilots. The SPLA also claimed a government F-5 fighter, though it may have crashed due to technical problems. In August 1986 it shot down a passenger plane using the SA-7.

Two years later the SPLA shot down a military helicopter and threatened air relief flights into the southern city of Juba. In the final three years of the first Sudanese Civil War 1955–72, Russia provided aircraft and most of the pilots for the Sudanese Air Force. Moscow's involvement was such that at one stage it was being called 'the Russians' Vietnam'. Although the southern rebels had no real air defences, in 1970 two Mi-8 helicopters were shot down and their Russian crews killed.

Civil war was also endemic in neighbouring Chad. As early as 1983, Libya lost a MiG-21 shot down by a SAM over Chad, followed by a Mirage four years later. Also, on 7 September 1987, French forces using Hawk missiles shot down a Libyan Tupolev-22 bomber attacking Chad's capital after the loss of the Libyan airbase at Maarten-as-Sarra. In 1983–84 French intervention cost two Jaguar jets destroyed and one Mirage and several helicopters damaged. One of the Jaguars was brought down by 23mm AAA fire.

Polisario in the Western Sahara began to deploy both the SA-6 and SA-7 against the Moroccan Air Force in the early 1980s. A Moroccan Mirage was shot down in mid-January 1985 by a missile launched from Algerian territory. In Ethiopia, the Eritrean separatists were not reported as receiving SA-7s until 1989, by which time the tide had already turned against the Ethiopian government.

During the early 1970s the Egyptians and Syrians, with Soviet assistance, constructed SAM networks even more formidable than those used by North Vietnam. The Arabs also deployed the SA-6 for the first time and it was this that posed the greatest threat to the Israeli Air Force. As it was fully mobile with unknown target acquisition radar frequencies, the Israelis were reduced to the expedient of dropping Second World War-style 'chaff' to blind it. The Israeli Air Force learned the hard way in the 1973 Yom Kippur War that before all else they must neutralise radar and SAM sites.

In eighteen days of fighting, the Israeli Air Force suffered, by its usual standards, appalling casualties – losing more than 25 per cent of its combat aircraft, mainly to radar-guided AAA rather than missiles (the Arabs accounted for 114 Israeli aircraft, of which the bulk were as a result of ground fire). For any other air force in the region, this would have been crippling. Crucially, the Israelis greatly benefited from America's experiences in the Vietnam War. The SA-2 and SA-3 used by the Egyptians were relatively immobile and most of their codes had been broken. Nor did the SA-6 threat last long either.

Air combat casualties were extremely low during the brief 1962 Sino-Indian conflict. This was largely due to the Indian government avoiding CAS for fear that the Chinese would retaliate against Indian cities. In the event, India lost just two helicopters to ground fire and abandoned a solitary helicopter and a transport plane. To add insult to injury, after the ceasefire China returned the helicopter in serviceable condition. China never issued any casualty figures. During the Suez crisis in 1956 the Israelis lost fifteen aircraft, eight of which were due to ground fire. Britain and France also lost ten planes to ground fire.

The Vietnam and Korean Wars saw the last major casualties incurred by aerial dogfights; in particular, the North Korean Air Force was simply torn from the sky. There was very limited air-to-air combat between Ethiopian Su-27s and Eritrean MiG-29s during the 1998–2000 border war. On four separate occasions the Su-27s, using Russian R-73 air-to-air missiles, proved superior, bringing down up to six MiG-29s. However, this was only half the rate achieved in the 1977–78 Ogaden War. In Europe, the Serbs lost six MiG-29Bs and one Mi-8 helicopter in dogfights during the 1999 NATO Kosovan air campaign. NATO efficiently destroyed 50 per cent of the Serbs' MiG-29 fleet in this manner. Previously, in 1995, USAF F-16s shot down four Bosnian Serb G-2 Galebs in the face of minimal aerial resistance.

During Desert Storm in 1991 the Iraqi Air Force lost thirty-five planes in air-to-air engagements. On 24 January 1991 a Saudi F-15 intercepted three Iraqi aircraft, shooting down two; from that point the Iraqi Air Force took little part in the war. The 1980–88 Iran–Iraq

War saw few air-to-air engagements and losses were never made public. During the Falklands War, Britain's only air-to-air loss was a Royal Marine Scout helicopter shot down by a Pucara. In contrast, thirty-one Argentine aircraft were lost in engagements with the venerable Harrier.

The Syrian Air Force lost ninety-one aircraft (including MiG-23s/25s) and six helicopters in only three days of aerial engagements with the Israelis in June 1982 over Lebanon's Bekaa valley. The Israeli Air Force, having learned its lesson in 1973, safeguarded its operations by first systematically destroying every single Syrian SAM site in the Bekaa. The American AIM-9L air-to-air missile was responsible for the bulk of the kills. Syrian losses would have been higher had Israeli pilots been allowed to pursue targets over Syria. The Israeli Air Force's achievement was impressive; according to USAF sources, approximately forty-four Syrian fighters were downed by Israeli F-16s, forty by F-15s and one by an F-4. Not a single Israeli warplane was lost in air-to-air combat.

In 1973 the Israeli Air Force lost a huge number of planes, though only fifteen were actually downed in air-to-air combat. The Egyptian–Israeli air war during the 1956 Suez crisis was also fairly limited, with only fourteen engagements recorded. The Israelis lost two aircraft in air-to-air combat and the Arabs seven. The disparity in Arab–Israeli air-to-air kills is glaring; in the past fifty years the Arabs have lost a total of 639 aircraft to this cause, the Israelis only thirty-nine.

Some of the few instances of air-to-air combat over Africa occurred in Angola on 30 September 1985, when the Angolan government claimed that eight South African Air Force planes had shot down six of its helicopters. They also bombed and strafed advancing Angolan government columns for three days running. On 16 July 1977 during the Ogaden War, two Ethiopian F-5As intercepted four Somali MiG-21MFs; two were shot down and the other two collided as they sought to evade an AIM-9B missile. By September 1977 the Somalis had lost twenty-three aircraft, ten of them in air-to-air combat. Between 1 and 4 April 1978, the Somali Air Corps claimed to have shot down three Ethiopian MiG-23s. This proved to be one of the highest air-to-air combat rates in Africa.

Apart from the Falklands conflict, there have been few instances of air-to-air combat in Latin America. One of these occurred during the brief war between El Salvador and Honduras. On 17 July 1969 Honduran fighters lost one aircraft but claimed three Salvadoran planes. The Honduran Air Force, which was considerably stronger, quickly defeated the Salvadoran Air Force. The Hondurans also forced down a Salvadoran Piper Cherokee and Salvador is believed to have lost at least another six fighters (it is not clear if these were shot down or caught on the ground).

During the 1965 Indo-Pakistan War the Pakistanis claimed thirty-six air-to-air kills, though Indian sources only acknowledge twelve. Pakistan recorded only eight aircraft lost in dogfights. Six years later in the 1971 Indo-Pakistan conflict, Pakistani claims of fifty-one kills were at variance with the Indian figure of eighteen to twenty-six. True losses remain difficult to ascertain because of ongoing tensions.

The first all-jet battle took place on 8 November 1950 when an American fighter took out the first Korean MiG-15. Despite the MiG-15's presence, the American F-86 Sabre achieved almost indisputable dominance over the skies of Korea, North and South. The US Far East Air Force had 350 combat planes at the start of hostilities and swiftly grappled air supremacy from the 110 North Korean piston-driven fighters. The first encounter occurred on 27 June 1950 when three North Korean Yak-3s fired on four US jets; all three were downed, followed by four more later in the day. The Soviet MiG-15 first appeared over Korea in late 1950, with some fifty aircraft flown by Chinese and Soviet pilots. The USAF claimed a total of 900 enemy aircraft, of which 792 were MiG-15s destroyed by Sabres, for the loss of only twenty-seven.

Aircraft remain at their most vulnerable whilst still on the ground, even when, as the Iraqis discovered, they are in concrete hardened shelters. Air defence and perimeter security remains paramount. In almost eighty years of aerial warfare about 770 aircraft have been destroyed on the ground. Most notably, Arab air forces have lost 566 aircraft to this cause. In most cases these have been caught by surprise air strikes.

In Operation Enduring Freedom, most of the Taliban Air Force was destroyed on the ground in the opening air strikes. The Ethiopians, in their raids on Asmara airport in 1998, tried unsuccessfully to catch the small Eritrean Air Force on the ground. Incapable of pressing home their attacks, they were driven off by the air defences. In the last air campaign conducted against Yugoslavia in 1999, forty-odd Serbian aircraft (including five MiG-29s and twenty MiG-21s) were likewise lost on the ground. During Desert Storm Iraq suffered 141 aircraft destroyed at its airbases, many of them by concrete-piercing precision-guided munitions.

Similarly, Ethiopia's Marxist government suffered most of its aircraft losses on the ground. The Somalis destroyed eight Ethiopian planes in air raids in July 1977. Eritrean separatists blew up about ten planes and helicopters in 1984 at Asmara airfield. They returned four years later and set fire to nine MiGs and a number of helicopters on 11 August 1988. The following year they destroyed another nine aircraft at Dubti, 250 miles north-east of Addis Ababa. The rest of the Ethiopian Air Force was largely destroyed in 1989 after leading a failed coup against President Mengistu.

During the Chadian Civil War of 1983–87 the Libyan Air Force suffered one of its greatest disasters without even taking off. Supporting Chadian rebels, Gaddafi had been conducting an air war using MiG-23s and Tu-22 bombers. In 1987 he lost twenty aircraft and a helicopter gunship in one swoop when Chadian government troops overran his airbase at Ouadi Doum inside Chad. The Chadian government also captured newly delivered SA-6 SAM systems. This was followed by the Libyan airbase at Maartan-As-Sarra inside Libya, where the Chadians destroyed another thirty aircraft. The war ended two years later with the Libyans humiliated. In the Sudan in 1985 the SPLA successfully attacked government airbases near Khartoum. In one raid they destroyed four MiGs and in another damaged three aircraft.

In 1982 Argentina lost thirty aircraft destroyed on the ground in the Falkland Islands. The Salvadoran Air Force suffered its worst losses when guerrillas attacked Ilopango air base on 27 January 1982; eight aircraft and six helicopters were completely destroyed and another four aircraft and a helicopter were severely damaged. However, the Americans made good the losses. Two years later a

Salvadoran Fairchild C-123 was destroyed by a mine whilst landing at San Miguel.

In 1973, despite Egypt's surprise attack on the Sinai, the Egyptian Air Force still lost twenty-two aircraft on the ground. Pakistan also tried to emulate Israel's 1967 success with an air attack on India on 3 December 1971. The Pakistani Air Force launched a series of attacks, compromised by the lateness of the raids, conducted on too narrow a front, with insufficient depth, plus the Indian Air Force had already dispersed its aircraft. Foolishly, the Pakistani Air Force committed just 30 per cent of its 300 combat aircraft, in part due to serviceability. Remarkably, India allegedly only lost two aircraft on the ground to these strikes, compared to the thirty-seven lost in 1965. The result was that Pakistan claimed a total of eighty-one aircraft (although India only acknowledged the loss of seventy-one) for the whole war, while India claimed ninety-four, though other sources put it as low as forty.

In 1967 during the Six-Day War the Israeli Air Force's timing and co-ordination was superb, particularly in catching the Egyptian Air Force in the early morning. The Egyptians lost a staggering 304 aircraft in a single day. The blow was decisive; Egypt's Air Force could not support its forces in the Sinai, giving the Israelis a free hand. After securing air superiority over the Sinai, the outcome was never in doubt. Israel then turned its attentions on its other Arab neighbours, claiming another ninety-nine aircraft. The Israelis destroyed the Jordanian Air Force, damaged the Iraqi Air Force and inflicted such losses on the Syrian Air Force that it took no further part in the fighting. Israel's losses for that day were just nineteen aircraft.

For the whole of the conflict, Israel claimed 452 Arab aircraft for the loss of just forty-six.

In summary, since the Second World War, at a very conservative estimate at least 7,000 aircraft have been destroyed in military operations. Of these, around 2,900 have been identified as being lost to ground fire and at least 800 destroyed on the ground, compared to 2,000 lost in air-to-air combat. In terms of global losses incurred during the numerous post-war conflicts, approximately 65 per cent were lost in the Asia–Pacific region (predominantly as a result of the Korean and Vietnam Wars) with more than 30 per cent lost in the Middle East (mainly as a result of the Arab–Israeli Wars). In the vast majority of these conflicts air defence forces constituted the biggest threat and took the greatest toll. Radar-guided AAA and SAMs remain an aircrew's greatest fear.

Air defence suppression is a discipline all of its own. Largely in its infancy during the Second World War, it has become a military science that armed forces around the world take very seriously. America in particular, after her ground-breaking experience in Vietnam, invested considerable resources in it and reaped the dividends over Iraq, Bosnia, Kosovo and Afghanistan. Likewise Israel, which learned the hard way over the Sinai, benefited over the Lebanon and relies upon it to give its air force an edge. Air Ops planners must consider the capabilities of an opponent's air defence forces and how to neutralise them, often even before considering the state of the opposition's air force. Whilst emergent technologies should help to reduce air war losses, in the face of such a continuing threat it seems the 30–70 ratio is set to stay for the foreseeable future.

23

KILLING HELICOPTERS

Operations Enduring Freedom and Iraqi Freedom provided a salutary reminder of how vulnerable combat
helicopters are to small arms fire. Over a decade earlier
things were completely different. In 1991, across the
sandy expanses of Iraq, hundreds of helicopters caused
an appropriate desert storm. America launched the largest helicopter air assault in military history; not since
Vietnam had such a vast heli-borne armada been gathered
in a single operation. In the opening stages of Operation
Desert Storm up to 450 helicopters were used to lift US
forces deep inside Iraq. Miraculously, none were reported
lost to enemy fire.

The helicopter's pre-eminence as a weapon of war was
dramatically reaffirmed in 1991. This was in stark contrast
to America's previous notable heli-borne foray into the
region with Operation Eagle Claw in April 1980, when five
helicopters were needlessly lost in Iran whilst conducting
an abortive rescue mission. However, since Desert Storm

America has not been able to replicate this dramatic success; in fact helicopter vulnerability has sometimes stymied operational deployment, such as in the case of the Balkans in the late 1990s.

Since the 1950s, approximately 6,600 military helicopters have been lost worldwide and their vulnerability has been highlighted by the intensity of the conflicts in which they have been involved. The largest helicopter combat losses occurred during the Vietnam War, the Soviet war in Afghanistan and the Iran–Iraq War. Indeed, the Asia–Pacific region has easily seen the largest loss of military helicopters, followed by the Middle East. In comparison, losses during conflicts notably in Africa, the Americas and Europe have been negligible.

Since it first appeared the increasingly sophisticated helicopter has borne the brunt of airlift, close air support (CAS), counter-insurgency (COIN) and medical evacuation (medevac) operations. Although helicopters have flown in support roles since the 1950s, it was not until the 1970s with the emergence of the gunship and anti-tank role that attack helicopters truly came into their own. The attack helicopter enjoys a number of advantages over fixed-wing CAS aircraft, most notably agility and low altitude. Unfortunately the latter makes it much more vulnerable to enemy ground fire. A correlation has emerged: the success of the helicopter as a weapon of war has driven the technological development of more vigorous countermeasures; in turn, ways have had to be found to improve the helicopter's battlefield survivability.

Whilst the helicopter first saw very limited service during the Second World War, it was in French Algeria in the 1950s that its potential as a transport, reconnaissance and casualty evacuation aircraft and as a gunship were first fully recognised. France's Algerian experience stood American forces in good stead for Vietnam, where the helicopter firmly established itself as the airborne assault weapon of the future.

Surprisingly, during Operation Enduring Freedom in Afghanistan the role of rotary wing aircraft was not as dramatic as might be expected. One of the first losses was a US Army CH-47 Chinook helicopter, which crash-landed in eastern Afghanistan on 9 January 2001. During the battles for Tora Bora in December 2001 an American strike force including helicopter gunships was used to attack a terrorist complex. US and Afghan government forces attacking the Shah-i-Kot valley during Operation Anaconda, launched on 2 March 2002, were supported by AH-64 Apache and AH-1W SuperCobra attack helicopters.

During this battle at least two Apaches were damaged by rocket propelled grenade (RPG) rounds, one of which was also discovered to have machine gun holes, two others were seriously damaged and a UH-60 was forced to land. Two MH-47 special operations helicopters carrying Navy SEALs were ambushed. Tragically, an RPG round seems to have cut the door gunner's tether, two more rounds struck the aircraft and a petty officer fell out as he tried to secure the gunner. The damaged MH-47 eventually had to be abandoned. Two additional MH-47s with a rescue force returned to the same landing zone and suffered another helicopter disabled. In contrast, the SuperCobras, which fight on the move, flew 217 sorties and seem to have escaped largely unscathed.

Russia suffered dramatic and significant helicopter casualties in Chechnya at the hands of Islamic militants (and more problematically mechanical failure) bent on ending Moscow's rule. It seems as if the Russian military forgot everything it had learned about helicopter warfare in Afghanistan during the 1980s.

As early as 5 June 1995, Chechen rebel fire brought down a Russian Mi-24 Hind (*Gorbach* or Hunchback) gunship near Nozhai Yurt. Later, a large Mi-26 transport helicopter fell from the sky on 19 August 2002, killing 118 people. Chechen militants are believed to have shot it down using a man-portable SAM. This was the second Mi-26 to be destroyed; the first crash-landed and caught fire on 24 September 1999. During 2001 Russian forces had two M-8 utility helicopters crash due to mechanical failure and one shot down by a Stinger SAM. The rebels also downed an Mi-24 gunship.

The year before Moscow lost at least five Mi-8s and one Mi-24. However, 1999 seems to have been by far the worst year with up to nine Mi-8s and one Mi-24 being lost through a combination of accidents and militant activity. The total number of Russian helicopters completely destroyed in Chechnya is difficult to ascertain. In 1999–2001 Western media reporting accounted for approximately fourteen helicopters, although this may well have included damaged as well as destroyed. Whatever the case, Moscow continued to suffer unacceptable losses under the disapproving glare of the Russian media.

In the Middle East the Israelis continued to deploy their US-supplied Apache and Cobra helicopters against the Palestinian Authority and suspected Hamas/Hezbollah terrorist targets. For example, on 24 January 2002, IDF/IAF AH-64 *Petens* (Python) were used to attack a car carrying senior Hamas terrorists, and on 7 February 2002, AH-64s attacked a similar Palestinian target in Nablus. Two days later they attacked two weapon workshops in the Gaza Strip, and such operations continued into 2004 (resulting in a mutiny by some pilots in October 2003 over collateral damage). The Palestinians do not seem to have come up with any effective countermeasures.

The Mi-24 gunship continued to be extremely popular in Africa with the region's resource-strapped militaries. For example, the Sierra Leone Air Force was operating several in 2001 against Revolutionary United Front (RUF) rebels until one crashed. Likewise, Burundi acquired three that were thought to have come from Ukraine, while during the Ethiopia–Eritrea conflict of 1998–2000 Ethiopia lost six Mi-24s brought down by Eritrean ground fire.

Fighting in the Democratic Republic of Congo (DRC, formerly Zaire) cost the Zimbabwean Air Force dearly. In the late 1990s Robert Mugabe's Zimbabwe obtained ten Mi-24s, which operated from Kariba, DRC. Russia modernised up to four in May 2000, making them night capable. However, Russian pilots and technical staff assigned to Thornhill air base were recalled after Zimbabwe failed to service its $35 million debt from the purchase of the helicopters. At least two are thought to have been lost in the DRC and only half of those remaining were service-able by mid-2001 when the Russians left. The supporting Namibian Defence Force (NDF) also lost two Indian-built Alouette helicopters in the DRC in 1998.

In the constant skirmishing with Pakistan over Siachen and Kargil during 1999–2000, India lost three Soviet-designed helicopters. During the 1999 NATO Kosovan air campaign the Serbs claimed to have shot down sixteen helicopters. In reality, America lost just two AH-64 Apaches and these were due to mechanical failure. The Serbs lost just one Mi-8 in a dogfight, as most of their helicopters, such as their Gazelle-GAMAs, were concealed during the air strikes and escaped largely unscathed. In post-Soviet Afghanistan from 1992 onwards an unknown number of helicopters and fixed-wing aircraft were either shot down or destroyed during the subsequent civil war. Estimates for losses by all causes in 1992–98 include about eighty transport helicopters and at least twelve gunships.

During the 1990s the Algerian government, struggling to contain the militant GIA (Armed Islamic Group, founded in 1992 by Algerian Afghan veterans), purchased thirty helicopters from Ukraine. The Algerian military's most notable success was in 1995 at Aïn Defla, when helicopter gunships, artillery, fighter-bombers and paratroops trapped 1,500 GIA fighters, killing at least half of them.

In 1991 the Coalition accounted for just six Iraqi helicopters; General Schwarzkopf had cause to regret that they did not destroy more. During the ceasefire talks on 3 March 1991, the Iraqis requested that in light of the damage done to their infrastructure they be allowed to move government officials around by helicopter. Without fully realising the consequences, Schwarzkopf agreed not to shoot down 'any' helicopters flying over Iraqi territory. Thus, using his helicopter gunships, Saddam Hussein was able to crush the Iraqi rebellion in Basra, Kabarla and the southern marshes with impunity, despite his defeat in Kuwait.

During the invasion of the tiny Caribbean island of Grenada in 1983 American losses were surprisingly high.

The rag-tag Grenadian and Cuban forces managed to shoot down or badly damage nine American Army and Marine helicopters (four UH-60 Blackhawks, two AH-1T Sea Cobras, one Hughes 500HD, one OH-58 Kiowa and one UH-46D Sea Knight) and cause minor damage to three others. In particular, two helicopter gunships were brought down by 23mm cannon fire.

Britain lost twenty-four helicopters during the 1982 Falklands conflict, fourteen of which were lost in operational accidents or when their parent ships were struck. Three Gazelles were shot down and one Royal Marine Scout was lost in air-to-air combat with an Argentine Pucara. The Argentines in fact lost a similar number, some twenty-six, mostly Argentine Army Pumas, including thirteen captured at Port Stanley. Those shot down mainly fell victim to the Harrier's 30mm cannon.

At least four Israeli helicopters were lost to Syrian ground fire and two to friendly fire during the 1982 war in the Lebanon. The year before, the Syrian Air Force lost two Mi-24s to Israeli aircraft whilst besieging Zahle in the Lebanon. During the Israeli invasion of Lebanon the Syrians lost at least seven helicopters (mainly SA.342M Gazelles) to ground fire and aircraft. One of them may have been shot down by an Israeli AH-1S Cobra using TOW anti-tank missiles and another was allegedly accounted for by an Israeli Merkava tank. Two Gazelles were also captured relatively intact and the Israelis were able to cannibalise one machine out of the two.

The helicopter's first major conventional war was Korea, where such aircraft as the Sikorsky S-51/H-5 and Bell 47 were used for casevac and reconnaissance purposes. Four helicopters accompanied the 1st Provisional US Marine Brigade to South Korea in 1950 and the first-ever heli evacuation of American casualties was performed by Marine VMO-6 helicopters on 4 August 1950. USMC HO-3 helicopters operated as battlefield support in the Changwon region. Losses on occasions were remarkably light: for example in September 1950 only one helicopter was lost during the daring Inchon landings.

The Marine helicopter squadron was used for every conceivable role and as early as 1951 the Marine Corps experimented with fitting assault helicopters with 2.75in rockets and machine guns. The American Army also attached bazookas to some of their helicopters during the Korean War. Ultimately the French were to lead the way with armed helicopters.

In the post-war period the helicopter was increasingly used in the colonial wars that beset the British and French Empires. In Indochina the French first deployed American-built H-19B helicopters against the Viet Minh (Vietnamese communists) in 1954. Regardless, it was too late for the French, who were defeated at Dien Bien Phu. France determined not to lose Algeria as well, and quickly became one of the world's leading authorities on the use of helicopters in combat conditions after buying further aircraft from the US. These included the Boeing vertol OH-21 Work Horse/Shawnee. In order to protect these slow transports from the Algerian National Liberation Front, the French Army's 2nd Helicopter Group experimented with mounting machine guns on its aircraft for COIN operations.

As a result of these experiences in the late 1950s, France pioneered the use of anti-tank helicopters for cave busting, which was to have resonance in 1990s Algeria. The French Army's Aviation Lègère de l'Observation Artillerie experimented mounting the SS.10 missile on their Alouette helicopters. A few were deployed in Algeria, where they were used against guerrilla hideouts. The French also used helicopters to protect the Morice Line in Algeria, designed to prevent the guerrillas crossing from Tunisia.

During the Malayan campaign of 1948–60 helicopters such as the RAF's Dragonfly HC.2 (S-51) and the Navy's Whirlwind (S-55) were first successfully used in COIN warfare. Helicopters were employed in medevac and inserting the SAS into remote jungle regions, although few were available and were mainly restricted to evacuation roles. In 1956 during the Suez crisis an ambitious British heli-borne landing some 3km beyond Port Said was to secure the Raswa Bridges. In the event only a few marines actually flew ashore; this was, though, Britain's first use of helicopters for assault purposes against a conventional enemy. Britain also deployed helicopters such as the Bristol Belvedere and Westland Wessex in Borneo and Aden during the 1960s.

Whilst the helicopter was used in assault operations in Korea, Suez and Algeria, the concept of massed helicopter air mobility was not fully appreciated until American intervention in Southeast Asia. The Vietnam War was where the helicopter truly came into its own, playing a significant role in the bloody and protracted conflict. Although the helicopter proved its utility in a whole host of roles, the war also highlighted its extreme vulnerability to small arms fire and ground fire in general. This was attested by the very high helicopter casualty rates, which have never been matched since.

Apart from the Huey, the Bell AH-1G Huey Cobra gunship, which went operational in September 1967, also made its presence felt during this war. In the case of the Cobra it was the first time a rotary wing aircraft was specifically designed for armed combat. Veterans of the Korean War, the Hiller UH-23 Raven and Bell OH-13 Sioux, also saw service in Vietnam, although neither was as successful as the light Hughes OH-6A Cayuse. The CH-53A had a rather unique role in that its primary mission was to rescue crashed helicopters. During January–May 1967 they retrieved 103 aircraft, seventy-two of them CH-34s. Only the Cobra proved to be particularly resilient; between September 1967 and June 1969, 563 AH-1s were hit by ground fire, but only fifty-seven were actually destroyed.

By 1968 the South Vietnamese Air Force possessed about seventy-five H-34 helicopters, but by the end of 1972 it had some 500 new machines; one of the largest, costliest and most modern helicopter fleets in the world. Just three years later, with the South's collapse, all were destroyed or had fallen into North Vietnamese hands. During the evacuation operations of 28–29 April 1975, so many South Vietnamese Air Force helicopters flew to the amphibious command ship USS *Blue Ridge* that many had to be pushed overboard. In the subsequent Sino-Vietnamese War of 1979 neither side made significant use of air power, although the Vietnamese undoubtedly made some use of their Hueys, quantities of which were seized at Tan Son Nhut air base in 1975.

Despite the loss of almost 7,000 helicopters in various conflicts around the world, the helicopter is considered an indispensable piece of military hardware. The problem of battlefield survivability will never be completely solved, but the scale of losses shows that casualties can be kept to a minimum and we are unlikely to ever see destruction on the scale of Vietnam and the Middle East again.

24

RISE OF THE DRONE

The beginning of the twenty-first century was character-ised by the rise of the armed drone, most notably the American Predator and Reaper unmanned aerial vehicles. Remotely piloted, they are capable of delivering missiles with pinpoint accuracy to very remote corners of the world. In the War on Terror they offered Washington greater freedom in tracking down and eliminating terror-ists on foreign soil. Their use led to the CIA's so-called 'Drone Wars' in which Washington authorised extra-judicial killings in countries with which America was not at war. The development of the armed drone was driven by 9/11 and the hunt for terror leader Osama bin Laden (who was eventually killed by US Special Forces in Pakistan on 2 May 2011).

After two weeks of close surveillance in Yemen, Operation Troy came to fruition on the morning of 30 September 2011 with the death of Anwar al-Awlaki, bin Laden's heir and the inspiration behind many of the acts and attempted acts of terrorism against the West. The drone attack reportedly took place near Khasaf, in Jawf province, 87 miles east of the Yemeni capital, Sana'a. The Drone Wars had claimed their latest victim.

This is certainly not the first time that US drones had targeted al-Qaeda leaders in Yemen. On 3 November 2002, Salim Sinan al-Harithi, also known as Abu Ali, who was one of those responsible for the USS *Cole* bombing, was killed in such a manner. This was the first time an armed drone had been used outside Afghanistan. Ahmed Hijazi, a US citizen, was also killed in the attack, which was another first.

Anwar al-Awlaki's white Toyota Hillux pick-up truck was just departing from a meeting with a tribal leader in northern Yemen when a Predator delivered its deadly payload. Two Hellfire anti-tank missiles were launched and the second hit the vehicle, ending the life of the man dubbed the 'Bin Laden of the internet'. Also killed were Samir Khan and three other senior associates. US-born Khan was the editor of *Inspire*, the highly effective internet site founded by al-Awlaki to recruit Islamic militants. It was believed that Saudi bomb maker Ibrahim Hassan al-Asiri may also have been killed in the strike.

Following the death of bin Laden there were only two real possible successors for the leadership of al-Qaeda. While Egyptian Dr Ayaman al-Zawahiri, long considered bin Laden's deputy, stepped into the role, al-Awlaki was a very viable rival for the job. He was a US citizen of Pakistani origin who had lived in Britain before moving to Yemen. Until his death he was a key figure with al-Qaeda in the Arabian Peninsula, which had operated from Yemen after being driven from Saudi Arabia.

Although al-Zawahiri served with bin Laden and was leader of Egyptian Islamic jihad, al-Awlaki in fact wielded far greater influence. His greatest skill was as an ideologue, propagandist and talent scout who fully appreciated the power of the internet as a recruiting tool for disaffected young Muslims. He was an imam at the mosques frequented by the 9/11 hijackers in America. He also influenced the British 7/7 bombers in 2005, the failed underpants bomber and the Fort Hood attack in 2009, and the attempted East Midland ink cartridge bombs and stabbing of British Member of Parliament Stephen Timms, both in 2010. His reach was wide and insidious.

US President Obama had signed the assassination order of al-Awlaki in January 2010. That year Yemeni commandos almost cornered him in a village in southern Yemen, but he slipped the noose. Then in May 2011 a US drone narrowly missed him when it attacked a convoy in which he was travelling. *Inspire* claimed that on that occasion US drones fired almost a dozen missiles.

While the beleaguered President Saleh of Yemen attempted to take the credit for Operation Troy, in order to bolster his regime with Washington, it was in fact a Saudi tip-off that alerted US intelligence to al-Awlaki's whereabouts. His death made Britain and America safer places.

After the East African US Embassy bombings in 1998 Washington offered $5 million for bin Laden's capture, as well as striking his camps in Afghanistan with Tomahawk cruise missiles. While America sought to capture or kill Osama bin Laden, opportunities were passed up in late 1998 and early 1999 to launch strikes on known locations where he was staying. Ironically, as early as December 1998 Bill Clinton had received a 'President's

Daily Briefing' entitled 'Bin Laden is Preparing to Hijack US Aircraft and Other Attacks'.

An unarmed Predator located bin Laden at the end of 1999 near Khost in eastern Afghanistan. Michael Scheuer, former CIA officer in charge of tracking bin Laden throughout the 1990s, recalled, 'We had no doubt over his identity. Bin Laden can clearly be seen standing out from the rest of the group next to the buildings. Nobody at the top of the CIA wanted to take the decision to arm Predator. It meant that even if we could find him we were not allowed to kill him.'

Clinton's options, though, were severely limited at that stage as the Predator was not armed, which meant sea-launched cruise missiles were all that could be deployed at short notice, but the attack on the Khalden training camp in the east of Afghanistan the previous year did not produce encouraging results. Nevertheless, a strike was planned for 11 February 1999 but information that visitors from the United Arab Emirates were present stalled the attack.

Dissatisfied with the CIA's poor results, Admiral Scott Frey, Director of Operations for the Joint Staffs, and Charlie Allen, in charge of collection priorities for the US Intelligence Community, came up with a novel way to hunt Osama bin Laden. Why not, they argued, use Predators to regularly loiter over Afghanistan to provide a real-time video feed on terror suspects. Unfortunately, Predator was then in short supply, being tied up over Bosnia and Iraq.

The attack on the *USS Cole* off Yemen showed that Al-Qaeda's war was spreading and by September 2000 the necessary technical means were in place to support a drone counter-terrorism mission in Afghanistan.

The CIA prepared to operate Predator from its HQ in Langley, Virginia, via the military's Central Command or CENTCOM HQ in Tampa, Florida, and a clandestine base in Uzbekistan at Karshi-Khanabad code-named K2 (Pakistan would have been the ideal host nation, but a military coup made it impossible). Between September and December 2000 the CIA, US Air Force and an interagency operations team conducted fifteen unarmed Predator reconnaissance missions over Afghanistan.

In testimony given by CIA head George Tenet in 2004 he said, 'During two missions the Predator may have observed Osama bin Laden. In one case this was an after-the-fact judgement. In the other, sources indicated that bin Laden would likely be at his Tarnak Farms facility, and, so cued, the Predator flew over the facility the next day.' This was a missed opportunity as Tenet notes, 'It imaged a tall man dressed in white robes with a physical and operational signature fitting bin Laden. A group of 10 people gathered around him were apparently paying their respects for a minute or two.'

Former senior White House official Richard A. Clarke was convinced that on at least three occasions in late 2000 Predator found bin Laden. Unfortunately the US Navy had no attack submarines on station, so Tomahawk missile strikes could not be launched. To make matters worse, the onset of winter in 2000 made it impossible to fly and the Predators were returned to America.

To get around the problem of not having cruise missiles or aircraft readily available, it was suggested using an armed version of Predator. It would be easier and less risky to deploy than men on the ground and a lot quieter than roaring jet fighter-bombers; in effect, with its long loiter time the drone provided an ideal silent assassin. The USAF had been working on mounting the AGM-114 Hellfire laser-guided, fire and forget air-to-surface missile, essentially an anti-tank weapon carried by the Cobra and Apache attack helicopters, on Predator. Its in-service date was scheduled for 2004 but this was dramatically accelerated after discussions between the USAF and the CIA.

Weapons tests were conducted between 22 May and 7 June 2001, but with mixed results. At the same time, in an effort to foster interagency co-ordination, two exercises were held to examine command and control issues and rules of engagement. Dubbed the MQ-1, the combination of Hellfire, Multi-spectral Targeting System and Predator seemed a winner.

In April 2001 Richard Clarke briefed the Bush administration's Deputies Committee, headed by the new Deputy National Security Advisor Steve Hadley. Clarke stated, 'We need to target bin Laden and his leadership by reinitiating flights of the Predator.' Ironically Paul Wolfowitz, Deputy Secretary of Defense, did not grasp the significance of the man he described as 'this little terrorist'. The administration did not fly drones over Afghanistan during its first eight months and was still refining the plan to employ the MQ-1 to assassinate al-Qaeda's leadership when the 11 September 2001 attacks on New York and Washington took place.

Washington could not decide whether to redeploy the unarmed Predator in early summer 2001 while the weather was good, or to wait for the armed variant. George Tenet observed, 'Some CIA officers believed that continued reconnaissance operations would undercut later armed operations.' The worry was that the Taliban would detect the flights, making Predator vulnerable to interception, anti-aircraft fire or surface-to-air missiles. Tenet elaborates, 'Additionally, indications were that the host country would be unlikely to tolerate extensive operations, especially after the Taliban became aware, as it surely would, of that country's assistance to the United States.'

In the meantime the key players could simply not agree if armed Predator should be used against those responsible for the numerous previous terrorist attacks. The CIA was concerned it could lead to reprisals against its operatives around the world. George Tenet was arguing a week before 9/11 that it would be a terrible mistake and it has been alleged that the CIA did not act sooner because the Directorate of Operations was risk adverse. After 9/11 there were understandably no objections to armed Predator operating as soon as possible over Afghanistan.

In fact, the CIA was authorised to deploy the Predator in early September 2001, but only on reconnaissance missions. Uzbekistan, which had not agreed to allow armed flights, frustratingly held up the delivery of the Hellfire missiles, no doubt while the thorny issue of appropriate rent for K2 was discussed. Just five days after 9/11 the missiles arrived, but the first armed flight did not occur until 7 October 2001 when host nation approval was finally granted, by which time Afghanistan was under general air attack as the Taliban had refused to hand over bin Laden. In the meantime the CIA had to make do with reconnaissance flights, which resumed over Kabul and Kandahar on 18 September.

During September and October 2001 there were at least four unsuccessful attempts to ambush bin Laden on the ground. The CIA-guided Predator got off to a very good start, killing al-Qaeda's No. 3, Mohammed Atef (alias Abu Hafs), in November 2001 near Kabul. By this stage two Predators had been lost to icing, leaving six-teen in total, though only one was in the air at any one time. The following month, acting on a tip-off, a Hellfire fired from Predator took out a Range Rover believed to be carrying bin Laden. DNA samples from the body later proved otherwise. This was probably the nearest Predator ever got to killing him before he escaped from Tora Bora and into Pakistan.

On 4 February 2002, a Predator tracked a group of up to twenty people apparently converging for a meeting and targeted six suspected al-Qaeda leaders (who it was hoped included bin Laden or his chief lieutenant Ayman Zawahiri) near the Zawahr Kili caves. Although the missile killed at least two of the party, the casualties were found to not include any of al-Qaeda's top leadership. Three days later a Predator attacked and destroyed a convoy transport-ing suspected terrorists.

The Predator was not only used against terrorists. The CIA tried unsuccessfully to kill Afghan warlord Gulbuddin Hekmatyar, after he called for the killing of US troops, on 6 May 2002. A Predator launched its Hellfires into the Shegal Gorge near Kabul, wounding thirty people, but missed its intended target. Following the attempt on Hekmatyar, the CIA remained very cau-tious about killing terrorists using the MQ-1. While targets and the number of attacks remain highly classified, former CIA official Mike Scheuer has stated that between May 2002 and February 2005 Predator fired fewer than ten missiles. This was not for the lack of targets, but because of legal constraints.

American intelligence assets including U-2 spy planes and Predator were deployed to 'assist' the Pakistani armed forces. CIA covert ops got more and more prominent on the Pakistani side of the border. US intelligence officials confirmed that a missile fired from a CIA-guided Predator killed a senior al-Qaeda operative in Pakistan on 13 May 2005. Although the CIA refused to confirm or deny operational details, bomb maker Haitham al-Yemeni had been under close surveillance in case he provided a lead on the whereabouts of Osama bin Laden. After the capture in north-west Pakistan of Abu Faraj al-Libbi, bin Laden's third in command, American and Pakistani authorities were concerned al-Yemeni would go to ground, so decided to act. Contrary to US statements, Pakistan denied al-Yemeni was killed on its soil.

The deployment of Predator to Pakistan was not new. President Musharraf admitted that CIA agents and technical experts were based in his country, and in early 2004 Predator assisted Pakistani armed forces conducting extensive security operations in South Waziristan. Subsequently a well-known pro-Taliban tribal leader, Nek Mohammad, and five of his companions were killed in Waziristan on 17 June 2004 by a laser-guided missile that probably came from a Predator. Mohammad had been using his satellite phone just before the missile struck and locals say they saw a drone overhead. Three years later Predator was supplemented by the larger MQ-9 Reaper, also known as Predator B. This was also being deployed in Pakistan.

While the war on terror had reasonable success against the Taliban and al-Qaeda lower-level leadership, bin Laden and his very senior entourage remained at large. By 2009 CIA Director Michael Hayden assessed that he was probably hiding in north-west Pakistan. This proved only partially accurate as he was in north-east Pakistan in the city of Abbottabad. Predator and Reaper ops continued unabated, particularly in Pakistan and Yemen, bringing swift retribution to those terrorists foolish enough to lower their guard. The drone, though, is not invulnerable. America confirmed on 2 October 2017 that a Reaper had been shot down over Yemen.

In the future even larger and more powerful unmanned aircraft will be taking to the skies. Washington ploughed millions of dollars into the development of such an aircraft for the US Navy. Only cost inhibited its deployment. Numerous other countries have been conducting research into this field, so it is only a matter of time.

Drones were not only employed in a hunter-killer role in the air war against terror. Smaller, light ones equipped with cameras, such as the American Desert Hawk and Raven, became key tools for providing real-time battlefield surveillance during the decade-long wars fought in Afghanistan and Iraq. These were used to gather tactical intelligence as well as call in air and artillery strikes. They were supported by the larger American Shadow reconnaissance drone, which had originally been developed for naval use. Such systems greatly impeded the freedom of movement by Afghan and Iraqi insurgents. They learned to fear the roving all-seeing eye in the sky.

Recent conventional conflicts such as the Russian invasion of Ukraine, which commenced with the occupation of Crimea in 2014 followed by a full-scale invasion eight years later, show that expensive large military drones are vulnerable to electronic warfare and air defence systems. Their size makes them relatively easy to detect and shoot down. This was the fate of the Turkish armed Bayraktar supplied to Ukraine. As a result the Ukrainians' remaining Bayraktars were restricted to reconnaissance work and kept away from Russian air defences.

By 2023 combat trends in Ukraine indicated that cheap, small commercially, off-the-shelf drones can be just as effective and their low cost makes them much more expendable. Their size enables them to carry out intelligence-gathering and even ordnance delivery missions at very low altitude well below radar cover. Although vulnerable to jamming and small-arms fire, they can be easily and swiftly replaced. Furthermore, there is no need for their operators to try to retrieve them when lost.

Both sides increasingly resorted to the extensive use of drones after the Russian air force singularly failed to gain air supremacy in the opening stages of the invasion. This came as a surprise as it was widely anticipated that the Ukrainians would be rapidly overwhelmed both in the air and on the ground. The failure of the Russian air force was partly due to its very limited performance and rapid improvements in Ukrainian air defences. Many risk-averse Russian pilots resorted to firing their missiles from their own airspace due to the threat posed by man-portable surface-to-air missiles. Despite extravagant claims by either side, there were very few actual dogfights. Ground fire accounted for most of the air losses.

The widespread use of small drones for real-time surveillance has now helped to revolutionise battlefield intelligence. Ironically, aerial warfare having gone hi-tech for so long has increasingly resorted to low-tech solutions. The drone will also undoubtedly play a role in the future of aerial warfare in attack and defensive roles. It will be employed as a buddy to help protect inordinately expensive fighter aircraft and their highly trained pilots while on missions. At the same time, swarms of drones will be used to attack or distract manned aircraft. The drone wars are here to stay and evolving drone designs are likely to progressively dominate the modern battle space.

BATTLE OF BRITAIN BOMBER

HEINKEL HE 111

The main target of the Spitfire and Hurricane during the Battle of Britain was the Heinkel bomber. This medium bomber was not comparable to the heavy Lancaster and Flying Fortress and was the wrong plane for the job. It was a prime example of planning for future operational requirements and getting it seriously wrong. The first prototype of what became the Heinkel He 111 piloted by Flugkapitän Gerhard Nitschke flew on 24 February 1935 at Rostock-Marienehe. Three further prototypes followed, with the second and fourth designed as civilian versions and the third a bomber. The He 111C airliner and the He 111G transport aircraft entered service with Lufthansa in the late 1930s.

In the meantime, ten He 111A-0 military pre-production versions were built. This had a longer nose than the third prototype and was armed with three

MG 15 machine guns in the nose, dorsal and ventral positions. Trials with these aircraft showed them to be underpowered with inadequate handling. All ten were rejected by the Luftwaffe and subsequently sold to China. A fifth prototype flew in 1936 powered by two 746kW (1,000hp) Daimler Benz 600A engines and was much more promising. It was followed by the He 111B-1 and He 111B-2 production models.

Only a few 111D were built due to the diversion of the required engines to fighter construction. The subsequent 111E and 111F were powered by Junkers Jumo engines. The early models of the 111 had elliptical wings, whereas the 111F was the first to feature a straight leading edge. The He 111P model that appeared in mid-1939 was powered by two 858kW (1,150hp) Daimler Benz 601Aa engines. It also introduced a fully glazed asymmetric nose, with its offset ball turret, in place of the stepped-up cockpits of the earlier variants. Not many 111Ps were completed before production was switched to the He 111H.

The major production version proved to be the H series, which was built in many different variants. The initial H-0 and H-1 were basically the same as the earlier 111P-2s except for the installation of 753kW (1,010hp) Jumo 211A engines, replacing the DB 601. The He 111H-2, which appeared in the autumn of 1939, had Jumo 211A-3 engines and carried two additional machine guns, one in the nose and one in the ventral gondola. The H-3 was armoured and armed with a 20mm MG FF cannon and an MG 15 in the ventral gondola, two MG 15s in the nose, one mounted dorsally and guns in the beam positions. The H-3 and H-5 were later fitted with a nose-mounted device to ward off barrage balloon cables and were subsequently redesignated the H-8. They were later converted into gilder tugs as the H-8/R2.

Efforts to increase the bomb load resulted in the H-16, which was introduced into service in the autumn of 1942. Similar to the H-11, it could carry up to 3,250kg of bombs. This extra weight came at a price and the aircraft required R-Geräte rocket-assisted take-off equipment to get it off the ground. The H-18 was designed as a pathfinder with exhaust flame dampers to help conceal it during night-time operations.

The standard crew of the He 111 was five, consisting of the pilot, navigator/bombardier and three gunners, one of whom also served as the radio operator. The pilot was seated offset to port in the glazed nose section, with the navigator/bombardier sitting next to him on a folding seat for take-off and landing. During the bomb run the navigator/bombardier would lay on a pad in the extreme nose, in order to use the bombsight. He also operated the nose gun.

By the autumn of 1944, more than 7,000 He 111s had been built for the Luftwaffe. Spain also built a total of 236 111Hs, designated the CASA 2.111, both during and after the war; around 130 of these had Jumo 211F-2 engines, while the rest were powered by the Rolls-Royce Merlin 500-29.

The first operational Luftwaffe squadron was with Kampfgeschwader or Bomber Group 154 based at Fassberg, which received its first deliveries in 1936. Early the following year, thirty He 111B-1s were sent to join the German Condor Legion supporting Franco's nationalists in Spain. These bombers served with Kampfgruppe 88 alongside other German aircraft types including the Bf 109B, He 51 and Ju 52. The latter was used for bombing operations but proved vulnerable to the newly arrived Soviet fighters flying in support of the republicans.

Although a trials squadron was ready in July 1937, re-equipping KG88 with the 111B took time and the conversion of its three original squadrons was not completed until July 1938. The Condor Legion was equipped with sixty-seven Ju 52s and those that survived were gifted to the nationalists. To replace them in a bombing role, a total of ninety-seven He 111s were supplied to the Condor Legion, consisting of sixty-one 111Bs and thirty-six 111Es. In contrast, just thirty-two Dornier Do 17 bombers were sent to Spain.

The experiences of the Condor Legion convinced the Luftwaffe that instead of developing a long-range heavy bomber force for strategic operations, they would concentrate on building up a twin-engine medium bomber force that would support German ground forces, forming a key component of the Blitzkrieg. This meant that the He 111, Do 17 and 215 and the Ju 88 were all medium bombers not intended for attacking targets at long distances. This was a critical error when it came to fighting the prolonged air wars on both the Eastern and Western fronts.

DEFENDER OF THE REICH

FOCKE WULF FW 190

In contrast, the Focke Wulf Fw 190 was exactly the right aircraft for its designated roles. It is not generally known that the Fw 190 is viewed as a far superior fighter than the much more famous Messerschmitt Bf 109 – and for good reason. This highly adaptable aircraft served in an enormous variety of roles, including fighter, night fighter, fighter-bomber/ground attack, reconnaissance and torpedo bomber throughout much of the Second World War. It initially appeared as an interim fighter designed to complement the Bf 109, and the first production model, the Fw 190A-1, was deployed on occupation duties in France in the summer of 1941, from where it could threaten the British Isles.

This meant that the Fw 190 fighter did not go into service until after Hitler's attack on the Soviet Union in late June 1941. Early production Fw 190s were armed with four machine guns mounted in the upper fuselage and wing roots. This proved inadequate firepower and aircraft were retrofitted with heavier cannon in each outer wing. When it made its appearance in the autumn of 1941 it quickly established Luftwaffe superiority for some months. The following year it began to replace the Bf 109 in Western Europe but the latter continued to serve in a range of roles on the Eastern Front. The Fw 190 became known as the 'Butcher Bird' by its opponents, which says much for its killing power over other fighters.

It was developed in the autumn of 1937 after the Reichsluftfahrtministerium placed a contract. Two proposals were developed by Kurt Tank, one with a liquid-cooled Daimler-Benz DB 601 engine and the other with the then new air-cooled BMW 139 radial. The latter was chosen with OberIng R. Blaser leading the design team the following year.

In May 1939 the prototype Fw 190V1 was completed, consisting of a cantilevered low-wing monoplane of stressed skin construction. It first took to the air at Bremen on 1 June 1939. The V2 flew that October. A decision to replace the BMW 139 with the longer and heavier BMW 801 meant relocating the cockpit further aft. This made for a much more comfortable flight for the pilot with fumes and overheating greatly reduced. The subsequent V3 and V4 prototypes were abandoned but the V5 with the new engine was finished in early 1940.

The V5 had its wingspan increased by 1 metre, which although slowing the aircraft slightly gave a superior climb and made it more manoeuvrable. These were vital prerequisites for a successful dogfighter. When a pre-production batch of thirty Fw 190A-0s were built the initial nine had the original shorter wing. In February 1941 the first aircraft were handed over to Erprobungskommando 190 based at Rechlin-Rogenthin for evaluation. The following month Jagdgeschwader 26 began to prepare for the new fighter's entry into service with the Luftwaffe. Its one-piece rearward sliding canopy gave good all-round visibility and the pilots immediately took to their new aircraft.

The early production version was dubbed the Fw 190A-1, which had the longer wings, the BMW 801C engine and was armed with four 7.92mm MG 17 machine guns. The later proved wholly inadequate after 6/JG 26 clashed with RAF Spitfires on 27 September 1941. Improvements led to the Fw 190A-2 that had a longer span and heavier armament, which in turn was followed by the excellent A-3 fighter-bomber. These were supplemented by the A-4 and A-5 with some of the latter modified as two-seater trainers. The Fw 190A-6 featured a lighter wing and armament of four 20mm cannon. The A-7 appeared in late 1943 equipped with two cannon and two machine guns. The final A-8 had two machine guns and could carry up to four cannon, plus bombs and rockets.

The up-gunned Fw 190A-2 was fitted with an improved engine and armed with two MG 17s mounted above the engine and two 20mm FF cannon in the wing roots. These were often supplemented by two additional MG 17s in the outer wing panels. The Fw 190A-3 featured the 1,800hp (1,342kW) BMW 801Dg engine. It had the cannons moved to the outer wing panels, with faster-firing MG 151s being installed in the root positions. Sub-variants of this model included the Fw 190A-3/U1 and A-3/U3 close support aircraft and the A-3/U4 reconnaissance aircraft. The A-4 that appeared in 1942 began to include the MW-50 water-methanol injection booster, which increased engine output for short periods and cranked the top speed up to 416mph. To cope with the conditions in the Mediterranean, the A-4/Trop was equipped with filters to protect the engine from choking dust and sand, and it could carry a bomb.

Problems with the engine overheating led to the Fw 190A-5 in 1943. This had a new engine mounting, which moved it 15cm forward. It was built in numerous variants, which included the A-5/U2 that had just two machine guns but could take bomb racks and drop fuel tanks. It was also fitted with flame-dampers so that it could operate at night. The U3 could carry bombs both on the fuselage and on the wings. Fighter-bomber versions included the U6, U8 and U11. The U14 and U15 were torpedo bomber variants.

The Fw 190A-6 was introduced in June 1943, incorporating a lighter-weight wing that could take four 20mm MG 151/20 cannon and was the forerunner of the A-6/R1 with six 20mm cannon, R2 with two 30mm cannon, R3 with an additional cannon beneath each wing and the R6, the final A-6 variant armed with a 210mm WGr.21 rocket launch tube under each wing. The A-7 appeared in December 1943 but was only built in small numbers, with 13mm machine guns replacing the 7.92mm nose-mounted machine guns.

The A-8 had greatly increased internal fuel storage and included numerous variants similar to the A6 family. Notable amongst these was the R7 with an armoured cockpit and the R11 all-weather fighter with heated canopy and radio navigation.

Problems with the Fw 190A's engine at altitude led to the 190B and 190C. These, though, were abandoned in favour of the 190D fitted with the Junkers Jumo 213A inline engine. This required a longer nose and to compensate the fuselage had to be lengthened. The Fw 190D-9 was known as the 'long-nose 190' or 'Dora 9'. This was armed with two wing-mounted MG 151/20 cannon and two MG 131 guns above the engine. It also had the MW 50 injection to boost emergency power. The wings could take drop tanks or bombs. Later production models were fitted with the bubble canopies installed on the Fw 190F.

Efforts to up-gun the 190D came to nothing. Several D-9s were converted to D-10 standard by installing the Jumo 213C engine, which permitted a 30mm cannon to fire through the propeller shaft and spinner. It did not go into production. Similarly, just seven prototypes of the D-11 were built, which had Jumo 213F engines, and were armed with two MG 151/20 cannon in the wing roots and two Mk 108s in the outer wing panels. The D-12 and D-13 were essentially ground attack aircraft. The proposed Fw 190E, a joint fighter-reconnaissance aircraft, was abandoned.

The Fw 190D-9 was a very fine aircraft and considered by many Luftwaffe pilots to be their best fighter. However, it was not available in any great numbers until early 1945 and by this stage the Luftwaffe was greatly hampered by shortages of experienced pilots and aviation fuel. The 190D had been preceded into service by the F-1, which appeared in early 1943 and was similar to the A-4. Likewise, the F-2 was related to the A-5 and the F-3 was similar to the A-6. The F-9 was the last in the F series and was an alternative version of the F-8. The initial ground attack variant was the Fw 190F, followed by the Fw 190G-1 fighter-bomber derived from the Fw 190A-5 but with a much greater bomb load. The Fw 190G-2 and G-3 were essentially the same but were equipped with Messerschmitt and Focke Wulf wing racks respectively.

More than 20,000 Fw 190s of all types were produced, compared to about 30,000 Bf 109s. The wide-track undercarriage of the Fw 190 meant that it was much more forgiving in the hands of an inexperienced pilot than the Bf 109. Also as a result it was much better than the Bf 109 at operating from rough airfields. This made it ideal for the Eastern Front, however there were never enough available and most were kept back to defend the skies over the Third Reich.

The Fw 190 served in North Africa from late 1942 until the German collapse in Tunisia the following year. It made its combat debut there on 16 November 1942, providing ground support. Over Tunisia it engaged American, British and French fighter aircraft, with its pilots achieving a high number of kills. The A-5 introduced in early 1943 proved an excellent fighter. Its range, speed and manoeuvrability were notably superior to the Bf 109. However, over the Eastern Front the Soviet Lavochkin La-5 was faster than both.

BOMBER COMMAND'S HEAVIES

SHORT STIRLING

Unlike the Luftwaffe, the RAF invested in developing heavy bombers that gave it strategic reach. Every schoolboy has heard of the iconic Lancaster bomber, but this was not the only type in Bomber Command's massive inventory. Yet apart from the much less glamorous Halifax, the Lancaster's other stablemates remain little known. In fact, the RAF started the war with three four-engine heavy bombers and half a dozen twin-engine medium bombers, many of which were deployed in a wide variety of roles. Getting the weight to power ratio right proved problematic with many of the early designs.

The first of the heavies was the Short Stirling, which failed to make its mark because it suffered from a very poor operational ceiling caused by a reduction in its wingspan. This was due to a penny-pinching edict from the Air Ministry that the aircraft had to fit inside existing hangars.

Not only was it the first four-engine heavy bomber to enter service, it was the only British four-engine bomber designed as such, as both the Lancaster and Halifax were developments of twin-engine designs.

On the ground the Stirling had a very pronounced nose-up position thanks to its very high landing gear. This tall undercarriage was designed to shorten take-offs and landings and featured the largest wheels fitted to an operational aircraft during the war. The nose-up attitude provided an adequate angle for sufficient lift during take-off. However, this made the aircraft difficult to handle on the ground. To compound matters, the undercarriage retraction motors could be unreliable and were not really up to the job.

The Stirling did not get off to a terribly auspicious start. It was designed to a 1936 requirement, but the prototype did not take to the air for another two years and then it was only half scale. The full-scale prototype flew on the eve of the Second World War in May 1939. On this inaugural fight the undercarriage gave way and the prototype was damaged beyond repair. Despite this, production was green-lighted.

The Stirling Mk I was powered by four engines that gave it a maximum speed of 270mph and a range of more than 2,000 miles. Capable of carrying 14,000lb of bombs, for self-defence the aircraft was armed with six .303in machine guns. Although unwieldy on the ground the Stirling proved manoeuvrable in the air and stable in flight. Nonetheless the 99ft wingspan plus a full bomb load meant the aircraft struggled to reach 12,000ft – in theory it was capable of 17,000ft. This left it vulnerable to enemy fighters and flak.

274 THE CHANGING FACE OF AERIAL WARFARE

The first unit to be equipped with the Stirling was No. 7 Squadron in August 1940. Initially based at RAF Leeming, in Yorkshire, it relocated to RAF Oakington, near Cambridge, in October 1940. It remained there for the rest of the war but converted to the Lancaster. The squadron conducted its first operational sortie on 10/11 February 1941, attacking an oil storage depot at Rotterdam.

The principal Stirling variant was the Mk III fitted with Hercules XVI engines and a twin gun dorsal turret. Despite all its shortcomings the Stirling equipped fifteen Bomber Command squadrons and served as a bomber until September 1944. Although no longer suitable as a bomber, the Mk IV and V were used as transports and glider tugs. In total 2,374 Stirlings were produced.

AVRO LANCASTER

Unlike many of Bomber Command's other bombers, the designers got it right from the start with the Avro Lancaster heavy bomber. It saw few modifications during its long production run. The prototype Lancaster Mk I flew on 9 January 1941 and the first operational sortie occurred on 2 March 1942 when four Lancasters dropped mines in the Heligoland Bight.

The Mk I was powered by four 1,223kW (1,640hp) Rolls-Royce Merlin XX, 22 or 24 12-cylinder Vee engines. This gave the aircraft a maximum speed of 287mph, an operational ceiling of 19,000ft and a range of 1,730 miles. Capable of carrying a 12,000lb bomb load, the Mk I (later renamed B.I and finally B.X) gained a reputation for being sturdy, pleasing to handle with good offensive and defensive firepower.

The most famous Lancaster raid was carried out by No. 617 Squadron in May 1943 against the Ruhr Dams. This operation employed 19 B.I (Specials) with a 'cut out' bomb bay to take the Barnes Wallis mine (known as the 'bouncing bomb'). Also, to save weight the nose and dorsal turrets were removed and in the tail turret the normal armament of four machine guns was reduced to two.

On the night of 15/16 September 1943 the squadron made the first operational use of the 12,000lb Tallboy bomb. These were used to attack the Dortmund–Ems canal. Later, on 14 March 1945, the enormous 22,000lb Grand Slam bomb was dropped by Specials on the Bielefeld viaduct. By the end of the war forty of these bombs had been dropped.

The shortage of Rolls-Royce Merlin engines resulted in the Lancaster Mk III (later B.III and B.3) powered by the American licence-built Packard V-1650 in its Merlin 28, 38 or 224 forms. It was also selected for production by Victory Aircraft Ltd of Toronto, Canada, which built 300 as the Mk X. In total 7,379 Lancasters were produced, including 3,294 Mk Is.

HANDLEY PAGE HALIFAX

Although much less famous than the Lancaster, the Handley Page Halifax served Bomber Command well over Germany. The first two prototypes flew in 1939 and initial production aircraft were dubbed the Halifax Mk I Series I. This was followed by the heavier Mk I Series II and the increased range Series III.

The main modification came with the Mk II Series I, which featured a two-gun dorsal turret and uprated 1,037kW (1,390hp) XX engines. The Mk II Series I (Special) was fitted with a fairing instead of the nose turret and the engine exhaust muff were omitted. The Mk II Series IA had a drag-reducing moulded Perspex nose that was standard on all subsequent Halifaxes, a four-gun dorsal turret and Merlin 22 engines. It also had rectangular vertical tail surfaces to reduce problems experienced with the original design.

The Mk III was powered by four 1,204kW (1,615hp) Bristol Hercules XVI radial engines in 1943. This gave it 282mph at 13,500ft, though the service ceiling was 24,000ft and a range of almost 2,000 miles. It was capable of carrying an internal bomb load of 14,500lb. Between the cockpit and the dorsal turret was a teardrop fairing that housed the direction finder aerial.

The dorsal turret was a Boulton Paul A Mk III mid-upper turret armed with four 7.62mm machine guns. The rear turret on the late models either had two or four machine guns. These also had on the underside of the fuselage a large radome for the H2S ground-mapping radar. It had originally been intended that the Halifax and Lancaster would have a ventral turret in this position. Other variants included the Mk VI and Mk VII, which also served as bombers. The Mk VIII and IX were transports.

BOMBER COMMAND'S MEDIUM BOMBERS

AVRO MANCHESTER

Most of the RAF's numerous early medium bombers were not successful. One, though, in particular proved to be an important stepping stone. Under specification P.13/36 the Air Ministry issued a remit for a twin-engine medium bomber, which resulted in the Avro Manchester. Although the airframe design was robust, especially the strong wing structure, power proved a problem. Fighter Command's Hurricanes and Spitfires needed the Merlin engines, so the Manchester by default used the 1,312kW (1,760hp) Rolls-Royce Vulture 24-cylinder engine. These were prone to catching fire and proved unreliable, making the Manchester a potential death-trap.

The Manchester prototype took to the air on 25 July 1939, with a second aircraft conducting trials on

26 May 1940. It was soon discovered that airflow along the fuselage was disrupted when the nose turret rotated whilst in flight. This was remedied by moving the axis of rotation forward slightly. The wingspan was also increased by 10ft.

The Manchester Mk I featured a distinctive central tail fin as well as twin fins and rudders. The engines gave a speed of 265mph at 17,000ft and the type could carry 10,350lb of bombs. The first twenty aircraft were followed by 180 Mk IAs with the central fin omitted. The Manchester became operational in November 1940 with 207 Squadron. Six aircraft took part in a raid on Brest harbour on the night of 24/25 February 1941. The upper gunner found the Frazer-Nash FN7 upper gun turret very uncomfortable, especially on prolonged flights.

The Rolls-Royce Vulture engines regularly failed to provide full power at vital moments. This and unexpected engine fires resulted in a loss rate of 40 per cent of Manchesters on operation and 25 per cent during training. As a result, the bomber was withdrawn from frontline service in 1942. However, the Manchester, apart from its troublesome engines, was a promising bomber and the Mk III was modified to take not two but four Merlin engines and the wingspan was increased from 90ft to 102ft. This became the prototype of the much more successful Avro Lancaster.

HANDLEY PAGE HAMPDEN

Besides the Manchester, the RAF's other main medium bombers are also worth mentioning. The Handley Page Hampden, while a reasonably good design, was hampered by a narrow fuselage, which confined the crew. The prototype first flew in June 1937 and delivery of the Mk I started in September 1938. In total 1,270 were constructed by two British manufacturers, with the Canadians building another 160. Another 100 aircraft were produced as the very similar Hereford training aircraft. The Hampden was powered by two 746kW (1,000hp) Bristol Pegasus XVII nine-cylinder, single-row radial engines. These provided 255mph, with a range of 1,885 miles and a service ceiling of 22,700ft, and they were capable of carrying 2,000lb of bombs. Initially used as a reconnaissance aircraft, the Hampden was found to be vulnerable to agile enemy fighters and was switched to being a night bomber before being withdrawn. It ended its days as a torpedo bomber conducting anti-shipping missions for Coastal Command.

ARMSTRONG WHITWORTH WHITLEY

The Armstrong Whitworth Whitley suffered the same fate as the Hampden. It entered service in 1937 and was obsolescent by the outbreak of war. Intended as a night bomber, it avoided the early heavy losses. However, it could not maintain altitude on a single engine and was withdrawn from front-line service in 1942. Nevertheless, more than 1,800 were built and they saw extensive action.

The Armstrong Whitworth Albemarle, which entered service in 1943, although intended as a medium bomber was only used in a transport/tug role.

VICKERS WELLINGTON

Lastly, the Wellington was one of the most important British bombers at the start of the war. It carried out most of the early operations, especially against the German fleet, until large numbers of heavies were available toward the end of 1941. It first appeared in October 1938 as the Wellington Mk I with 726kW (1,000hp) Pegasus XVIII radial engines. The Mk III had Rolls-Royce Merlin Vee engines, the IV used Pratt & Whitney Twin Wasp radials, the VI had Merlins and the X used Hercules. Notably, Wellingtons formed the major part of the very first 1,000 bomber raid. It also had a second career in maritime reconnaissance, transport and training roles. By October 1945 some 11,461 had been delivered.

APPENDIX 5

NORMANDY TANK BUSTER

HAWKER TYPHOON

The Typhoon was the epitome of an aircraft being designed for a task that it proved wholly unsuited for and instead excelling at something else. It came into being in the 1930s intended as a replacement for the Hurricane fighter aircraft. Its role was to be a medium- and high-altitude interceptor designed to counter Hitler's Luftwaffe. As with so much British military equipment designed and manufactured during the Second World War, it was clearly a rushed job, suffering critical engine and structural problems in its early stages. In fact, it proved so problematic that the RAF almost abandoned it altogether.

The Typhoon first flew in early 1940, but it would be another two years before it became operational due to constant teething problems, principally the engine jamming or violent engine and tail vibration causing fatal structural failure. Initially there was talk of making it a night fighter, but the exhausts were in the pilot's eye line. A day fighter role was also suggested but although it was designed as an all-altitude interceptor it soon proved inadequate in this role except at low level.

The first two prototypes powered by a 1,566kW (2,100hp) Napier Sabre H-type 24-cylinder inline liquid-cooled sleeve-valve engine conducted their test flights on 24 February 1940. The first production model did not appear until the following year. It was soon found that the engine sleeves had a very nasty habit of jamming, causing the cylinder to explode. Additionally, violent engine and tail vibration could cause structural failure and sudden deceleration could cause violent yaw.

Other problems encountered included poor visibility due to the canopy struts (this was eventually solved by a bubble hood) and dangerous levels of carbon monoxide in the cockpit. Nor could the pilot see behind due to the armour plating. Pilots found the cockpit hood and door appalling; the doors opened outward and the windows wound down with car door handles (these were not altered until 1943).

Typhoon production was centred at the Gloster Aircraft Company Ltd, Brockworth, Gloucester. A single batch of fifteen aircraft was also produced by Hawker at Langley, being intended as trials aircraft, however some were delivered to the RAF. By 1941 the RAF had about 150 Typhoons and it was now they discovered that the tail had a nasty habit of falling off. In December 1941 Squadron Leader Dundas went to Fighter Command calling for the teething problems to be solved, particularly the blind rear view. Hawker's designer Sydney Camm was understandably displeased at being told by a 21-year-old that the

Typhoon's blind spot was a bad design. A heated row followed, with Camm arguing its shortcomings were negated by the aircraft's speed. Nonetheless, all the aircraft were sent back for modification.

Throughout late 1941 to mid-1943 tests were conducted at the Aeroplane and Armament Experimental Establishment, Boscombe Down. Worryingly, the aircraft continued to suffer problems well into 1943 and these were not rectified until the end of the year in time for D-Day. Following bomb load tests the Typhoon was cleared to carry two 500lb bombs. In addition, the Mk 1B was fitted with the more reliable 1,625kW (2,180hp) Sabre IIA engine. This gave it a speed of 412mph (663kmh) with a ceiling of 35,200ft and a range of 980 miles (1,577km). Speed wise it easily outperformed the Hurricane, which managed in excess of 300mph, and could give the Spitfire Mk XIV ground attack aircraft, which could manage 448mph, a run for its money. More importantly, its speed outstripped the Messerschmitt Bf 109 and even the Focke Wulf Fw 190.

The Mk 1A was armed with 12 x 7.7mm machine guns, but in the Mk 1B these were replaced with powerful 4 x 20mm cannons, plus 8 x 60lb high-explosive rockets or 25lb armour-piercing rockets. Squadron Leader Roland Beamont of 609 Squadron conducted a series of ground attack sorties over Europe that highlighted its value in this role in support of Operation Overlord and the Normandy landings. It was this task that ultimately made it famous.

AIR COMBAT TRENDS 1948 TO PRESENT DAY

AIRCRAFT LOSSES BY CAUSE

Region	Air-to-air combat	Other
Africa	25	199
Americas	36	150
Asia–Pacific	1,150	3,181
Europe	12	45
Middle East	729	1,474
Total	**1,952**	**5,049**

AIRCRAFT LOST TO GROUND FIRE

Conflict	No. of aircraft
Afghanistan 2001	1
Kosovo 1999	1
DRC 1998–2001	9
Ethiopia-Eritrea 1998–2000	21
Bosnia 1995	1

Conflict	No. of aircraft
Iraq 1991	15
Sri Lanka 1983–89	9
Sudan 1983–89	4
Chad 1983–87	3
Falklands 1982	33
El Salvador 1981–89	1
Angola 1980–90	31
Nicaragua 1980–88	3
Afghanistan 1979–88	600
Arab–Israeli 1973	142
Indo-Pakistan 1971	36
Arab–Israeli 1967	37
Indo-Pakistan 1965	37
Sino-Indian 1962	2
Vietnam 1962–72	1,530
Arab–Israeli 1956	13
Sudan 1955–72	2
Korea 1950–53	340
Arab–Israeli 1948–49	29
Total	**2,900**

AIRCRAFT DESTROYED ON THE GROUND

Conflict	No. of aircraft
Afghanistan 2001	20
Kosovo 1995	40
Iraq 1991	141
Sudan 1983–89	7
Chad 1983–87	22
Falklands 1982	30

Conflict	No. of aircraft
El Salvador 1981–89	20
Ethiopia 1977–90	18
Ogaden 1977–78	8
Arab–Israeli 1973	22
Indo-Pakistan 1971	4
Arab–Israeli 1967	373
Indo-Pakistan 1965	37
Sino-Indian 1962	1
Arab–Israeli 1948–49	30
Total	**773**

AFRICAN CONFLICTS – AIRCRAFT LOSSES BY CAUSE

Conflict	Air-to-air	Other
Ethiopia 1998–2000	6	22
DRC 1998	0	18
Sudan 1983–89	0	13
Chad 1983–87	0	49
Angola 1980–90	6	31
Ethiopia 1977–90	0	28
Ogaden 1977–78	13	24
Rhodesia 1965–79	0	2
Nigeria 1967–70	0	10
Sudan 1955–72	0	2
Total	**25**	**199**

AFRICAN CONFLICTS – AIRCRAFT LOSSES BY COUNTRY

Conflict	Country	Losses	Country	Losses	Total
Ethiopia–Eritrea War 1998–2000	Ethiopia	21	Eritrea	7	28
DRC 1998	Angola, Namibia and DRC	18	Rebels	0	18
2nd Sudanese Civil War 1983–89	Sudanese Government	13	SPLA separatists	0	13
Chadian Civil War 1983–87	France	3	Libya	46	49
Angolan Civil War 1980–90	Angolan Government	37	UNITA separatists	0	37
Ethiopian Civil War 1977–90	Ethiopian Government	28	Separatists	0	28
Ogaden War 1977–78	Ethiopia	14	Somalia	23	37
Rhodesia 1964–79	Rhodesian Government	2	Guerrillas	0	2
Nigerian Civil War 1967– 70	Nigerian Government	0	Biafran separatists	10	10
1st Sudanese Civil War 1955–72	Sudanese Government	2	Southern rebels	0	2
Total					**224**

AMERICAS CONFLICTS – AIRCRAFT LOSSES BY CAUSE

Conflict	Air-to-air	Other
Falklands 1982	32	119
El Salvador 1981–89	0	21
Nicaragua 1980–88	0	3
El Salvador–Honduras 1969	4	7
Total	**36**	**150**

AMERICAS CONFLICTS – AIRCRAFT LOSSES BY COUNTRY

Conflict	Country	Losses	Country	Losses	Total
Falklands 1982	Argentina	117	Britain	34	151
El Salvador 1981–89	Salvadoran Government	21	FMLN guerrillas	0	21
Nicaragua 1980–88	Nicaraguan Government	2	Contra guerrillas	1	3
El Salvador–Honduras 1969	El Salvador	10	Honduras	1	11
Total					186

ASIA–PACIFIC CONFLICTS – AIRCRAFT LOSSES BY CAUSE

Conflict	Air-to-air	Other
Afghanistan 2001	0	23
Sri Lanka 1983–2001	0	12
Indo-Pakistan 1995–99	0	4
Afghanistan 1979–88	0	100
Indo-Pakistan 1971	26	85
Indo-Pakistan 1965	20	74
Vietnam 1963–72	285	2,165
Sino-Indian 1962	0	4
Korea 1950–53	819	714
Total	1,150	3,181

ASIA–PACIFIC CONFLICTS – AIRCRAFT LOSSES BY COUNTRY

Conflict	Country	Losses	Country	Losses	Total
Afghanistan 2001	USA	22	Taliban	20	23
Sri Lanka 1983–2001	Sri Lanka	12	Tamil Tigers	0	12

Conflict	Country	Losses	Country	Losses	Total
Indo-Pakistan Siachen and Kargil 1995–99	India	4	Pakistan	0	4
Afghanistan 1979–88	Russia	100	Mujahideen	0	100
Indo-Pakistan 1971	India	71	Pakistan	40	111
Indo-Pakistan 1965	India	75	Pakistan	19	94
Vietnam 1963–72	North Vietnam	193	USA	2,257	2,450
Sino-Indian 1962	China	0	India	4	4
Korea 1950–53	North Korea	900	UN	633	1,533
Total					**4,330**

EUROPEAN CONFLICTS – AIRCRAFT LOSSES BY CAUSE

Conflict	Air-to-air	Other
Kosovo 1999	8	44
Bosnia 1995	4	1
Total	**12**	**45**

EUROPEAN CONFLICTS – AIRCRAFT LOSSES BY COUNTRY

Conflict	Country	Losses	Country	Losses	Total
Kosovo 1999	NATO	5	Serbia	47	52
Bosnia 1995	NATO/UN	1	Serbia	4	5
Total					**57**

MIDDLE EAST CONFLICTS — AIRCRAFT LOSSES BY CAUSE

Conflict	Air-to-air	Other
Gulf War 1991	35	313
Lebanon 1982–98	85	19
Iran–Iraq 1980–88	100	300
Iraqi–Kurdish War 1974–75	0	2
Yom Kippur 1973	292	264
War of Attrition 1968–70	117	63
Six-Day War 1967	72	426
Suez Israeli 1956	9	25
Arab–Israeli 1948–49	19	62
Total	**729**	**1,474**

MIDDLE EAST CONFLICTS — AIRCRAFT LOSSES BY COUNTRY

Conflict	Country	Losses	Country	Losses	Total
Gulf War 1991	Coalition	50	Iraq	298	348
Lebanon 1982	Israel	7	Syria	97	104
Iran–Iraq 1980–88	Iran	150	Iraq	250	400
Iraqi–Kurdish War 1974–75	Iraq	2	Kurds/Iran	0	2
Yom Kippur 1973	Egypt and Syria	442	Israel	114	556
War of Attrition 1968–70	Arab	150	Israel	30	180
Six-Day War 1967	Egypt, Iraq and Jordan	452	Israel	46	498
Suez 1956	Egypt	9	Britain, France and Israel	25	34
Arab–Israeli 1948–49	Arab	56	Israel	25	81
Total					**2,203**

ESTIMATES OF GLOBAL HELICOPTER COMBAT LOSSES 1955 TO PRESENT DAY

Africa	57
America	69
Asia–Pacific	5,846
Europe	26
Middle East	586
Total	**6,584**

EUROPEAN CONFLICTS – HELICOPTER LOSSES BY COUNTRY

Europe	Country	Losses	Country	Losses	Total
Chechnya 1999–2002	Russia	23	Chechen militants	0	23
Kosovo 1999	NATO	2	Serbia	1	3
Subtotal					**26**

AFRICAN CONFLICTS – HELICOPTER LOSSES BY COUNTRY

Conflict	Country	Losses	Country	Losses	Total
DRC 1998–2000	Zimbabwe and Namibia	4	Rebels	0	4
Ethiopia–Eritrea War 1998–2000	Ethiopia	6	Eritrea	0	6
Somalia 1993	America	2	Somali warlords	0	2
Second Sudanese Civil War 1983–89	Sudan	4	SPLA separatists	0	4
Chadian Civil War 1983–87	Chad	0	Libya	3	3

Conflict	Country	Losses	Country	Losses	Total
Angolan Civil War 1980–90	Angola	13	UNITA separatists	0	13
Libya 1986	Libya	3	America	0	3
Ethiopian Civil War 1977–90	Ethiopia	18	Separatists	0	18
Rhodesian Civil War 1964–79	Rhodesia	2	Guerrillas	0	2
First Sudanese Civil War 1955–72	Sudan	2	Separatists	0	2
Subtotal					**57**

ASIA–PACIFIC CONFLICTS – HELICOPTER LOSSES BY COUNTRY

Conflict	Country	Losses	Country	Losses	Total
Afghanistan 2001–2002	USA	3	Taliban	0	3
Sri Lanka 1983–2001	Sri Lanka	6	Tamil Tigers	0	6
Indo-Pakistan Siachen and Kargil 1995–99	India	3	Pakistan	0	3
Afghanistan 1992–98	All factions	92	/	0	92
Afghanistan 1979–89	Russia	500	Mujahideen	0	500
Indo-Pakistan 1971	India	4	Pakistan	0	4
Indo-Pakistan 1965	India	75	Pakistan	19	94
Vietnam 1962–75	South Vietnam	500	America	4,642	5,142
Sino-Indian 1962	China	0	India	2	2
Subtotal					**5,846**

MIDDLE EAST CONFLICTS – HELICOPTER LOSSES BY COUNTRY

Conflict	Country	Losses	Country	Losses	Total
2nd Gulf War 2003	America and Britain	12	Iraq	0	12
1st Gulf War 1991	America	21	Iraq	6	27
Lebanon 1982	Israel	6	Syria	9	15
Iran–Iraq 1980–88	Iran	250	Iraq	250	500
Yom Kippur 1973	Egypt	16	Israel	0	16
Six-Day War 1967	Egypt, Iraq and Jordan	16	Israel	0	16
Subtotal					**586**

AMERICAS CONFLICTS – HELICOPTER LOSSES BY COUNTRY

Conflict	Country	Losses	Country	Losses	Total
El Salvador 1981–89	El Salvador	8	FMLN guerrillas	0	8
Nicaragua 1980–88	Nicaragua	2	Contra guerrillas	0	2
Grenada 1983	America	9	Grenada	0	9
Falklands 1982	Argentina	26	Britain	24	50
Subtotal					**69**

BIBLIOGRAPHY

Air Ministry. *Bomber Command Continues: The Air Ministry Account of the Rising Offensive against Germany July 1941–June 1942*. London: His Majesty's Stationery Office, 1942.

Arthur, Max. *Forgotten Voices of the Second World War*. London: Ebury Press, 2004.

Bekker, Cajus. *The Luftwaffe War Diaries*. London: Corgi, 1972.

Bishop, Patrick. *Battle of Britain*. London: Quercus, 2010.

Bishop, Patrick. *Bomber Boys: Fighting Back 1940–1945*. London: Harper Perennial, 2008.

Bishop, Patrick. *Fighter Boys: Saving Britain 1940*. London: Harper Perennial, 2004.

Borovik, Artyom. *The Hidden War: The True Story of War in Afghanistan*. London: Faber and Faber, 1991.

Bowman, Martin W. & Boiten, Theo. *Raiders of the Reich: Air Battle Western Europe: 1942–45*. Shrewsbury: Airlife Publishing, 2003.

Bowyer, Chaz. *History of the RAF*. London: Hamlyn, 1979.

Boyd, Alexander. *The Soviet Air Force since 1918*. London: Macdonald and Jane's, 1977.

Boyne, Walter J. & Handleman, Philip (ed). *Brassey's Air Combat Reader: Historic Feats and Aviation Legends*. Washington DC: Potomac, 2005.

Braithwaite, Roderic. *Afghantsy: The Russians in Afghanistan 1979–80*. London: Profile, 2011.

Braybrook, Roy. *Battle for the Falklands (3) Air Forces*. London: Osprey, 1982.

Brickhill, Paul. *Reach for the Sky*. London: Fontana, 1965.

Brookes, Andrew. *Air War over Russia*. Hersham: Ian Allan, 2003.

Catchpole, Brian. *The Korean War*. London: Constable, 2000.

Chant, Chris. *Aircraft of World War II: 300 of the World's Greatest Aircraft 1939–1945*. London: Amber, 2016.

Chant, Chris. *Air War in the Gulf 1991*. Oxford: Osprey, 2001.

Collier, Richard. *Eagle Day: The Battle of Britain August 6 –September 15 1940*. London: J.M. Dent & Sons, 1980.

Cooper, Alan. *Bravery Awards for Aerial Combat*. Barnsley: Pen & Sword Aviation, 2007.

Corbin, Jimmy. *Last of the Ten Fighter Boys*. Stroud: The History Press, 2010.

Deighton, Len. *Fighter: The True Story of the Battle of Britain*. London: Jonathan Cape, 1978.

Dewar, Lieutenant-Colonel Michael. *The British Army in Northern Ireland*. London: Arms and Armour Press, 1985.

Doe, Helen. *Fighter Pilot*. Stroud: Amberley, 2016.

Donald, David (ed). *The Pocket Guide to Military Aircraft and the World's Air Forces*. Twickenham: Temple, 1986.

Dorr, Robert F. *Air War South Vietnam*. London: Arms and Armour, 1990.

Embry, Air Chief Marshal Sir Basil. *Mission Completed*. London: Methuen, 1957.

Ethell, Jeffrey & Price, Alfred. *Air War South Pacific*. London: Sidgwick and Jackson, 1983.

Feifer, George. *Okinawa 1945: The Stalingrad of the Pacific*. Stroud: Tempus, 2005.

Francillon, René. *Vietnam Air Wars*. Twickenham: Aerospace, 1987.

Franks, Norman. *Battle of Britain*. London: Bison, 1981.

Franks, Norman. *RAF Fighter Command 1936–1968*. Sparkford: Patrick Stephens, 1992.

Franks, Norman. *Typhoon Attack*. London: Grub Street, 2003.

Gunston, Bill. *An Illustrated Guide to Military Helicopters*. London: Salamander, 1981.

Gunston, Bill. *An Illustrated Guide to Modern Airborne Missiles*. London: Salamander, 1983.

Gunston, Bill. *An Illustrated Guide to Modern Bombers*. London: Salamander, 1988.

Harris, Marshal of the RAF Sir Arthur. *Bomber Offensive*. London: Collins, 1947.

Hastings, Max. *Bomber Command*. London: Pan, 1999.

Holmes, Tony (ed). *Dogfight: The Greatest Air Duels of World War II*. Oxford: Osprey, 2013.

Hyams, Jacky. *Spitfire Stories*. London: Michael O'Mara, 2017.

Infield, Glenn B. *The Poltava Affair: The secret World War II operation that foreshadowed the Cold War*. London: Robert Hale, 1974.

Irving, David. *The Rise and Fall of the Luftwaffe: The Life of Luftwaffe Marshal Erhard Milch*. London: Weidenfeld and Nicolson, 1973.

Isby, David. *Russia's War in Afghanistan*. London: Osprey, 1986.

Jackson, Robert. *Aircraft of World War II: Development, Weaponry, Specifications*. Enderby: Silverdale, 2005.

Jackson, Robert. *Before the Storm: The Story of Bomber Command 1939–42*. London: Cassell, 2001.

Jackson, Robert. *Bomber! Famous Bomber Missions of World War II*. London: Arthur Barker, 1980.

Johnstone, Air Vice Marshal Sandy. *Spitfire into War*. London: Grafton, 1988.

Kent, Johnny. *One of the Few*. Stroud: The History Press, 2017.

Knight, Dennis. *Harvest of Messerschmitts: The Chronicle of a Village at War 1940*. London: Frederick Warne, 1981.

Lewis, Damien. *Apache Dawn*. London: Sphere, 2009.

Levine, Joshua. *Forgotten Voices of the Blitz and the Battle of Britain*. London: Ebury Press, 2007.

Macy, Ed. *Apache*. London: Harper Press, 2008.

Macy, Ed. *Hellfire*. London: Harper Press, 2009.

Mason, Robert. *Chickenhawk*. London: Corgi, 1984.

Mondey, David. *Axis Aircraft of World War II*. London: Bounty, 2006.

Mosley, Leonard. *The Battle of Britain*. Chicago: Time Life, 1984.

Neil, Wing Commander Tom. *Gun Button to Fire*. Stroud: Amberley, 2011.

Overy, Richard. *The Battle of Britain*. London: Penguin, 2004.

Peters, Flight Lieutenant John & Nicol, Flight Lieutenant John. *Tornado Down*. London: Signet, 1993.

Ray, John. *The Battle of Britain: Dowding and the First Victory, 1940*. London: Cassell, 2002.

Richardson, Nick. *No Escape Zone*. London: Little, Brown, 2001.

Richey, Paul. *Fighter Pilot*. Stroud: The History Press, 2016.

Ripley, Tim. *Conflict in the Balkans 1991–2000*. Oxford: Osprey, 2001.

Sarkar, Dilip. *The Few: The Story of the Battle of Britain in the Words of the Pilots*. Stroud: Amberley, 2012.

Sarkar, Dilip, (ed). *Air Ministry Spitfire Manual 1940*. Stroud: Amberley 2010.

Spector, Ronald H. *Eagle Against the Sun: The American War with Japan*. London: Cassell, 2000.

Spick, Mike & Ripley, Tim. *The New Illustrated Guide to Modern Attack Aircraft*. London: Salamander, 1992.

Sweetman, John. *Bomber Crew: Taking on the Reich*. London: Abacus, 2006.

Sweetman, John. *Schweinfurt Disaster in the Skies*. London: Pan/Ballantine, 1971.

Taylor, Frederick. *Dresden Tuesday 13 February 1945*. London: Bloomsbury, 2005.

Tolland, John. *Infamy: Pearl Harbor and its Aftermath*. London: Penguin, 2001.

Verrier, Anthony. *The Bomber Offensive*. London: Pan, 1974.

Wilson, Kevin. *Bomber Boys: The Rhur, the Dambusters and Bloody Berlin*. London: Cassell, 2006.

Wright, Derrick. *Pacific Victory: Tarawa to Okinawa 1943–1945*. Stroud: Sutton, 2005.

Wynn, Humphrey & Young, Susan. *Prelude to Overlord*. Shrewsbury: Airlife Publishing, 1983.

INDEX